Mathematical Success for Students on the Margin

Voices, Practice, and Promise

Thea K. Dunn

University Press of America,® Inc.
Lanham · Boulder · New York · Toronto · Oxford

4501 Forbes Boulevard
Suite 200
Lanham, Maryland 20706
UPA Acquisitions Department (301) 459-3366

PO Box 317
Oxford
OX2 9RU, UK

Printed in the United States of America
British Library Cataloging in Publication Information Available

Library of Congress Control Number: 2005932821
ISBN 978-0-7618-3333-8

™ The paper used in this publication meets the minimum requirements of American National Standard for Information Sciences—Permanence of Paper for Printed Library Materials, ANSI Z39.48—1984

To my family, with much love and appreciation

Contents

Preface

Chapter One	Transforming Teaching and Learning	1
Chapter Two	Theory and Perspectives	8
Chapter Three	Methods of Inquiry and Analysis	24
Chapter Four	The Preservice Teachers	30
Chapter Five	The Learners	106
Chapter Six:	Conclusion	148
Bibliography		151
Index		163

Preface

This book evolved out of a commitment to improve mathematics teaching and learning for students on the margin and to begin to reverse the cycle of educational failure for students labeled "at-risk." Although research in teacher preparation has explored the ways in which preservice teachers learn to teach mathematics, few studies have focused on how teaching marginalized students interacts with and complicates this process. Paralleling the need to improve the preparation of preservice teachers to work with marginalized students is the need for more effective mathematics education programs for students who are academically at risk. Therefore, this book investigates the influences of a pre-student teaching experience in an alternative high school on preservice teachers' beliefs and instructional practices. The journeys of the learners provide the reader with a window on their perceptions and mathematical development and serve as catalysts for critically reflecting on the mathematics education of marginalized students. In addition to presenting case studies and testimonies, this book offers discussion and exploration prompts at the end of each chapter. The voices of preservice teachers and marginalized learners are heard in this book and encourage the reader to think critically about mathematics education from the perspective of teacher and learner. By exploring the influences of the preservice teachers' beliefs, perceptions, and instructional strategies on the mathematical development of marginalized students, this book invites reflection on the relationship between teacher preparation activities and the mathematical education of students on the margin.

Thea Dunn
Bergheim, Texas

Chapter One

Transforming Teaching and Learning

> Because mathematics hold the key to leadership in our information-based society, the widening gap between those who are mathematically literate and those who are not coincides, to a frightening degree, with racial and economic categories. We are at risk of becoming a divided nation in which knowledge of mathematics supports a productive, technologically powerful elite while a dependent, semiliterate majority find economic and political power beyond research. Unless corrected, innumeracy and illiteracy will drive America apart. (National Research Council 1989, 14)

Equity in the mathematics classroom may be defined in three ways: (1) equal opportunity to learn mathematics; (2) equal educational treatment; and (3) equal educational outcomes (Fennema & Meyer, 1989). This letter interpretation moves beyond the traditional definition of equity as "equal opportunity" to one that supports equal achievement and future participation in mathematics and "ensures that students have a real chance to become engaged in and to learn from the academic core that they encounter" (Secada, 1989, p. 40). Resnick (1995) identifies the two challenges facing American education today as raising overall achievement levels and making opportunities for achievement more equitable. Recent mathematics education reforms provide vision and direction for school mathematics (CBMS, 2001; MSEB & NRC, 1990; NCTM, 2000). A prominent reform-oriented view advocates teaching mathematics in such a way that all students develop "mathematical power" (NCTM, 1989, 2000). Mathematical power includes both ability (to conjecture, reason logically, solve nonroutine problems, and communicate about mathematics) and attitude (self-confidence and a disposition to question, explore, and engage in significant mathematical problems). This goal requires that teachers use knowledge of how students construct conceptualizations of mathematical concepts to select and use materials, develop multiple teaching strategies and instructional activities, design and utilize authentic assessment strategies, and create learning environments in which all students learn mathematics meaningfully. Implementation of these reforms regarding mathematics curriculum and instructional practice requires that teachers and students engage in problem solving, reasoning, communicating, and making mathematical connections (Dunn, 2002).

Mathematics education reform documents articulate a vision of high expectations and make a conscious effort to enhance how learners experience mathematics. They view mathematics classrooms as communities of learners, encourage the use of cooperative problem solving groups, and support mathematical thinking and communication by creating supportive environments for all learners. However, in terms of equity mathematics education in the United States remains unevenly distributed (Strutchens & Silver, 2000). along lines of social class (Secada, 1992) and economic background (Tate, 1995). While reform efforts advocate high standards for all learners, current educational practices continue to relegate a disproportionate number of learners to remedial track classes (Oakes, 1990). By exaggerating the disparities between learners, the practice of tracking leads to differentiated outcomes. This educational differentiation further disadvantages already disadvantaged learners (Oakes, 1985). Although the United States professes to have a democratic educational system, inequitable curriculum and pedagogy sustain the gap between marginalized and mainstream learners. Reform efforts' emphasis on higher academic standards for learners has adversely affected "at-risk" learners, many of whom did not meet previous minimum standards (Waxman, 1992). According to Secada (1992), a broader conception of teaching, one that is not acknowledged by the reform efforts, takes into account pressing equity concerns. These concerns include learner access to advanced mathematics courses, perseverance in course taking, and, ultimately, participation in mathematics-related careers. It has been suggested that a useful strategy would be to "take the rhetoric of goals and standards and wrap it in the same package as the equity agenda" (Kozol, 1991, p. 18).

Instructional practices reflect teachers' conceptions which resonate with their own experiences and background (Thompson, 1984; Cooney, 1985; Ernest, 1991; Cabello & Burstein, 1995). Mathematics teachers' knowledge, conceptions, and attitudes about students and student learning impact the way in which these teachers interact with students in their classrooms (Calderhead, 1984). Because teachers act upon their expectations of students, negative teacher conceptions or low expectations for their students influence classroom practices and may adversely affect student performance (Strutchens, 2000; Brophy, 1985). Learners labeled as less capable than their peers are marginalized. That is, they are taught less mathematics and are presented with skill-oriented, direct instruction, and practice rather than conceptually-focused instruction promoting problem solving and understanding (Campbell & Langrall, 1993). Rote instruction often fails because it reinforces learners' negative self-perceptions and deprives them of cognitive stimulation (Silver, Smith, & Nelson, 1995).

Teachers and Teacher Preparation

The attitudes of teachers play an important role in influencing the interests, motivations, and aspirations of marginalized learners. Teachers engage in differential behaviors toward their students based upon race, gender, class, language and/or differential student ability (Grant & Sleeter, 1993). They also tend to develop perceptions and expectations of learners based upon race, ethnicity, gender, and social class (Secada, 1992). Teachers' acceptance of societal stereotypes about race, ethnicity, gender, social class and student ability shapes their behaviors toward and their expectations for learners. Like *Pygmalion in the Classroom* (Rosenthal and Jacobson, 1968), the expectations and belief systems of teachers are often reflected in their teaching behavior (Thompson, 1984; Thompson, 1992). In turn, that behavior impacts how learners respond and perform (Grant, 1989).

Inequitable teaching can occur within a single classroom when a teacher denies certain learners access to educational opportunities by providing stimulating content and instruction to a select group (Rist, 1970). Because the *a priori* knowledge, expectations, attitudes, and beliefs of teachers are included among the factors associated with differences in learners' performances (Secada, 1992), the occurrence of lowered expectations for educationally at-risk learners suggests negative implications for the effective education of these learners. Secada notes that, low-track mathematics courses, where lower socioeconomic status, minority, and other marginalized learners are often found, tend to be taught by beginning teachers or teachers who have fallen out of favor with school administrators. These courses, in addition to offering less mathematical content, do not cover content in depth or in ways that foster student engagement (Secada, 1992).

In an effort to remedy the disempowerment of marginalized learners, Cummins (1989) identifies five general principles for teacher education. In addition to adequate content knowledge, teacher preparation programs must prepare teachers to: (1) guide and facilitate rather than control student learning; (2) encourage students to work collaboratively in a learning context; (3) assist learners in establishing a genuine dialogue; (4) facilitate the development of learners' higher level cognitive skills; and, (5) plan, design, and implement tasks that generate intrinsic rather than extrinsic motivation. These principles imply that, if learner empowerment is to be attained, teacher education programs must prepare teachers to guide learners to achieve their highest potential in mathematics. However, according to Kagan (1992), teacher preparation courses fail to provide teachers with adequate knowledge of classrooms, adequate knowledge of learners, the extended practica needed to acquire that knowledge, or a realistic view of teaching in its full classroom or school context. She cites five components of professional growth among novice and beginning teachers: (1) an increased awareness of what they know and believe about learners and classrooms; (2) an acquisition of knowledge about learners and the use of this knowledge to reconstruct their personal images of self as teacher; (3) a shift in attention from self to the design of instruction to student learning; (4) the development of standard procedures that integrate instruction and management; and, (5) growth in problem solving skills. These components of professional growth suggest that, in order to meet the criteria outlined by Cummins (1989), teacher preparation programs must provide preservice mathematics teachers, not only with extended practica, but with practica that give preservice teachers the opportunity to explore and cultivate instructional strategies that facilitate learners' acquisition of conceptual understandings. A crucial gap in many novice teachers' knowledge concerns the nature of the learners (their abilities, interests, learning styles) and ways to design instruction to meet their needs (Grossman, 1990). This suggests that engaging preservice teachers in teaching experiences is crucial to the process of their acquiring knowledge of teaching and learners and utilizing this knowledge to modify, adapt, and reconstruct their self-images as teachers.

Students on the Margin

There are two groups of learners who have been overlooked by educational reform movements: (1) learners who have been alienated from economic and educational opportunities; and (2) learners who are marginalized because they cannot succeed in the problem-solving skills required in current school practices (Jones, 1988). Marginalized learners are excluded from full and productive participation in the classroom, do not obtain equal access to quality learning, and are at risk of not realizing their personal and academic promise. Learners who are at-risk in the current educational system are those who are not receiving an education that would allow them full or meaningful participation in the industrial/capitalistic social order or in the informational/technical culture of the twenty-first century (Eubanks and Parish, 1992). When students' home resources and experiences differ from the expectations on which school experiences are built (Levin, 1992), they are often marginalized and at risk of not realizing their personal and academic promise. Social and economic stratification, racism, and lack of access to and prior experiences with education contribute in powerful ways (Kozol, 1991; Mehan, 1991; Oakes, 1986) to the marginalization of students. Mathematics for these learners is typically perceived as a hierarchy of skills that are learned in a particular sequence (Carey, Fennema, Carpenter, & Franke, 1995).

Characteristics of the Educational Environment

Waxman (1992) notes that because the term "at-risk" focuses on individual characteristics, it labels and stigmatizes the learner. Further, the term suggests that it is the learner who has the characteristic of being at risk rather than being in an educational environment or circumstance that is considered to be at risk. An at-risk educational environment is one that (1) alienates learners; (2) provides low standards and a low quality of education; (3) has differential expectations for learners; (4) is unresponsive to learners; or (5) does not adequately prepare learners for the future (Sinclair & Ghory, 1987). While the literature suggests that learners are marginalized due to factors related to their socioeconomic status, family background, or community, it is more likely that learners become marginalized or disconnected because schools are not meeting their specific educational needs (Baptiste, 1992). While distasteful to many, the term "at-risk" persists because of its wide acceptance and research base (Walker de Felix, 1992).

Characteristics of the Learner's Environment

In addition to being at-risk because of their educational environment, at-risk learners have also been defined as those who lack the home and community resources to benefit from conventional schooling practices (Levin, 1987). They are also learners who have been exposed to inappropriate education within their family or community (Pallas, et al., 1989). Some of the factors that place students at risk are the circumstances of their lives, such as homelessness, poverty, or a dysfunctional family (McCown, Driscoll, & Roop, 1996). These problems manifest themselves in low self-esteem, poor problem solving skills, low achievement, high absenteeism, and dysfunctional behavior (Garcia & Walker de Felix, 1992). Despite adequate intellectual endowment, these learners are at risk of underachieving in school (Comer, 1988). Unless fundamental social and educational restructuring occurs, at-risk learners will continue to be filtered out of mathematics courses and excluded from mathematics-related careers.

Students on the margin may not have been taught the necessary life skills because their families lacked structure and support (Ogden & Germinario, 1988) and their home resources and experiences differ from the expectations on which school experiences are built (Levin, 1992). They often live in urban poverty or rural isolation, are below grade level in mathematics and reading skills, are low academic achievers, exhibit poor attendance or behavior problems, or are culturally or linguistically different from their peers (Slavin, 1989). In her discussion of the at-risk secondary mathematics student, Vatter (1992) summarizes the characteristics of at-risk learners as: (1) poor self-concept; (2) poor academic performance, high absenteeism, and discipline problems; (3) low aspirations and parents or guardians with low expectations; (4) low family socioeconomic level; (5) nontraditional family life, often with a single or foster parent, or with a stepparent; and (6) inadequate goals or lack of future orientation. Because marginality is related to the way a person is perceived and treated, Sinclair and Ghory (1987) suggest that educators need to: (1) create convincing reasons for their students to learn; (2) provide appropriate resources, including alternative curriculum and pedagogy; and, (3) place marginal learners in responsible roles that enhance their self-esteem, social behavior, and academic achievement.

Transforming Teaching and Learning

The initiative discussed in this book attempts to reverse the cycle of educational failure for marginalized and disconnected learners who have not flourished within traditional schools. This initiative primarily involves two components. The first is to improve preservice teachers' preparation to foster mathematics achievement among all students, including marginalized and disconnected students, by focusing on teacher beliefs, self-concept, instructional strategies, and perceptions of students on the margin. The second component of this initiative is to enhance the mathematical development and self-concept of marginalized and disconnected students.

A primary goal of this initiative, to produce instructional change and to improve the conditions for learning, developed out of a recognition that existing instructional practices underestimate and constrain marginalized learners' achievement (Moll & Diaz, 1987). Rather than addressing the low mathematics performance of marginalized learners, low-level mathematics courses contribute to the poor performance of these learners (Oakes, Ornseth, Bell, & Camp, 1990). A further goal of this initiative is to broaden the instructional focus from an emphasis on the acquisition of procedural skills toward one that incorporates the development of conceptual knowledge into instruction. In addition, this initiative provides teachers with information about at-risk learners that they can use in the classroom and to redesign mathematics curriculum and pedagogy.

This book explores the experiences of preservice teachers as they planned, designed, and implemented mathematical experiences for learners in an alternative high school. It presents the voices, visions, and perspectives of the marginalized learners as they engaged in problem-based mathematics lessons designed to promote mathematical ability and achievement. Specifically, it focuses on the preservice teachers' professional development as they engaged in supervised teaching experiences with high school mathematics learners labeled "at-risk." In addition, it explores the connections between teaching and learning, including the impact of the preservice teachers' instructional practices on student learning.

The pre-student teaching experience that provides the setting for this book has the potential to enhance instructional support for the education of marginalized students in mathematics. Further, it provides a model to support and enhance the development of preservice teachers as they attempt to design and implement instructional experiences intended to enhance the engagement and mathematics achievement of marginalized and disconnected learners who attend an alternative high school. Research and evaluation of this teaching experience provide data on the relationships between teacher preparation activities, instructional practices, and learning in an alternative high school which may be utilized to inform decisions about both preservice teacher preparation and the mathematical education of students on the margin (Dunn, 2001). Moreover, this teaching experience discussed here is designed to be replicated in order to enhance the preparation of preservice teachers and promote greater mathematical achievement by students on the margin.

For Exploration and Reflection

1. Reflect on your own mathematical journey as a student of mathematics. Discuss your experiences, beliefs, and development. How does engaging in this process assist you in developing an understanding of how a student's prior mathematics background and learning experiences influence mathematics learning, attitudes, and achievement?
2. Identify and describe different factors that put students at risk of being marginalized or disconnected in the mathematics classroom. Discuss your explanations with several teachers and/or principals and ask them to react to your explanations and to share their views about why certain learners may become marginalized. Document their contrasting views and note whether they believe students are at-risk because of individual or environmental characteristics.
3. Visit several mathematics classrooms and focus on observing the differences in the ways teachers respond to and interact with different students. Are there students on the margins who the teachers seem to ignore? Do students engage in the mathematics lessons in different ways? Describe. Discuss how you would respond to and support these students.

Chapter Two

Theory and Perspectives

The purpose of this chapter is to provide a context and framework within which to examine the issue of fostering mathematical success for students on the margin. In an effort to clarify the relationship between this issue and previous work conducted in related areas, theory and perspectives on mathematical knowledge, mathematical understanding, teacher knowledge, beliefs and attitudes, mathematical self-concept, and mathematical discourse provide a lens through which to view the themes discussed in this book.

Mathematical Knowledge

While a Platonist perspective assumes that mathematics expresses eternal, objective relationships between objects – a body of knowledge "out there," social constructivists view mathematics as a social construct involving values, orientations, processes and concepts - all of which are subject to agreement and refutation (Lerman, 1983). According to this perspective, mathematical knowledge is both a historical creation of social groups and a personal construction (Carraher, Schliemann, & Carraher, 1988). As such, it is shaped by both cultural and personal values. In contrast to the notion of mathematics as an absolute body of hierarchically organized facts, a humanistic view of mathematics reasons that mathematics is both socially constructed and fallible. Rather than being passively received, knowledge is actively built up and derived from experience (von Glasersfeld, 1990). A dynamic, inquiry-oriented view of mathematical knowledge translates into a perception of learning as a generative process of meaning making that is informed by both context and purpose (Borasi, 1996). Instruction emphasizes the critical role of social interactions in learning (Moll and Diaz, 1987). Learners are influenced by both personal and community constructions (Cobb, 1994). Similarly, teaching is conceived of as creating an environment that is conducive to stimulating and supporting learners' inquiry (Borasi, 1996). However, rather than viewing mathematical knowledge as a dynamic, social construction, teachers and students typically perceive of mathematics as an absolute, static, external body of knowledge that is hierarchically organized and context- and value-free (Cooney, 1985; Dossey, 1992). This dualistic perspective views mathematics as an absolute system in which results are uniquely determined and learners are either right or wrong. This perception of mathematical knowledge translates into a view of learning as the progressive accumulation of isolated bits of information and skills that are acquired by listening, observing, memorizing, and practicing. Similarly, conceiving of mathematical knowledge as a body of established facts and techniques translates into a view of teaching as the direct transmission of knowledge from teacher to learners (Borasi, 1996). Many educators discuss the nature of mathematical knowledge by focusing on the complementary notions of procedures and concepts (Lesh, Landau & Hamilton, 1984; Hiebert & Lefevre, 1986). A brief description of these kinds of mathematical knowledge follows.

Conceptual Knowledge

Conceptual knowledge reflects an understanding of principles, relations, and patterns. It refers to knowledge of the underlying structure of mathematics and is attained by integrating new knowledge with existing knowledge (NCTM, 2000). As such, conceptual knowledge of mathematics consists of an understanding of relationships that are integrated with or connected to other mathematical ideas and concepts. Because it is equated with connected networks, conceptual knowledge is rich in relationships (Hiebert & Lefevre, 1986). As networks of internal representations are constructed and restructured, learners create conceptual knowledge (Hiebert & Carpenter, 1992; Hiebert & Lefevre, 1986). The process of constructing conceptual knowledge can be guided or facilitated by the teacher (Hatano, 1988). Hatano offers several principles for mathematics education based on facilitating learners' constructing conceptual knowledge. These include posing interesting and novel problems; placing problems within pragmatic or familiar contexts; encouraging learners to bring in prior skills or ideas; using peer interactions; offering the use of tools such as calculators, mathematical symbols, and graphics; and providing ample opportunities for reflection.

Procedural Knowledge

While conceptual knowledge is commonly defined as knowledge that is understood, procedural knowledge is defined as a sequence of actions (Hiebert & Carpenter, 1992) and mastery of computational skills (Eisenhart, et al., 1993). Procedural knowledge is the ability to recognize symbols and follow rules to solve mathematical problems. It can be thought of as possessing facility or skill in carrying out sequential actions (Hiebert & Carpenter, 1992). While instruction that emphasizes memorizing and imitating may facilitate the acquisition of procedural knowledge (e.g., the ability to efficiently manipulate mathematical symbols to complete mathematical exercises), such strategies may actually thwart learning by interfering with the construction of conceptual knowledge (Anthony, 1996).

Relationship between Conceptual and Procedural Knowledge

According to Hiebert and Carpenter (1992), mathematical expertise involves both procedural and conceptual knowledge. It is generally agreed that learning mathematics involves a complex interaction of conceptual and procedural knowledge rather than the acquisition of one or the other (Hiebert & Wearne, 1986). Hatano (1988) posits that before learners are able to use a mathematical procedure, they must understand it, interpret it in relation to their prior knowledge, and reconstruct it in their mind. However, Hiebert and Carpenter (1992) argue that it is possible to solve any problem without understanding. Learners who are competent performing routine mathematical tasks often have idiosyncratic interpretations of the mathematical notations, weak conceptions, and significant misunderstandings of the mathematical concepts underlying those tasks (Erlwanger, 1975;Thompson, 1992). When students learn procedures without understanding, procedural knowledge consists of meaningless routines, and those procedures become dissociated from conceptual knowledge (Hiebert & Lefevre, 1986). In contrast, procedures connected to conceptual knowledge give flexibility to mathematical thinking and enable learners to extend the range of both types of knowledge when problem situations arise (NCTM, 2000). By considering a specific area of mathematical content, algebra, the relationships between procedural and conceptual knowledge can be illustrated more clearly. For most learners, the focus of their arithmetic activities has been manipulating symbols to find particular numerical answers. However, algebra requires learners to think in terms of the structure of the problems rather than in terms of the operations used to solve them. This involves deriving procedures, interpreting and understanding relationships, and expressing these in general, simplified form (Kieran, 1992). Although solving algebraic word problems requires both text processing and mathematical skills, algebra instruction is often procedural and focuses on the manipulation of mathematical expressions (Kaput and Sims-Knight, 1983). Algebra learners who overgeneralize previously learned arithmetic procedures to more advanced mathematical topics are likely to develop misconceptions that present obstacles to learning algebra (Matz, 1980).

Knowledge in Context

Mathematical knowledge acquisition is situated in contexts. Rather than being subordinate to or separable from learning, the situation is integral to what is learned, coproducing knowledge though activity (Brown, Collins & Duguid,

1989). The mathematical knowledge that learners or novice users have is closely tied to the context of its acquisition, such as how a new mathematical concept is introduced. Even when two learners each know a mathematical concept, how it is represented may vary among them. In contrast, the mathematical knowledge that experts or experienced users have is no longer closely associated with the situation in which it was originally acquired. This so because the knowledge of experts reflects not only its original acquisition but also the long history of its use in various situations (Hatano, 1988). Based upon their conceptual understandings, experts are able to adapt known procedures, or invent new procedures, to new situations. Unless the goal of instruction is to teach routine procedures that are useful only for limited problem solving, it is critical to develop learners' conceptual understandings of mathematical concepts. Traditional instructional practices, which emphasize the learning of standard algorithms, may result in learners developing instrumental beliefs about school mathematics. Moreover, these learners may construct relatively immature, syntactically-based understandings of mathematical concepts (Cobb & Wheatley, 1988). Instruction that emphasizes syntactic rules (routinized procedures) may actually interfere with the adoption of semantic analyses - connecting symbols with referents and developing rules (Wearne & Hiebert, 1988). At-risk learners are often tracked into classes in which the instructional focus is on low-level knowledge, remediation, the acquisition of routine procedures, the use of worksheets, and the decomposition of information into isolated pieces (Oakes, 1985). Learners in lower track classes are not exposed to mathematical opportunities that would nurture the acquisition of conceptual knowledge. According to Hatano (1996), although mathematical knowledge is not about the material world as such, it is useful for solving real-world problems in an accurate and efficient manner. For many people, mathematical knowledge is a set of tools by which they can solve real-world problems. That is, mathematics consists of procedures for converting real-world problems into mathematics representations and of procedures for manipulating these representations in order to find mathematical solutions.

Connecting Knowledge to Symbols

Three cognitive processes have been identified as contributing to the connecting of mathematical knowledge to mathematical symbols: representations, reflection, and automatization (Hiebert & Carpenter, 1992). Mathematical ideas are represented in real-life events through oral and written language and by objects, pictures, graphs, and symbols. Reflecting on what has been represented results in an awareness of relations and patterns and enables learners to construct mathematical meanings. Learners engage in critical thinking, make inferences, and develop abstract ideas as they make connections between and within representations. Automatization involves automatic or effortless performance. Automatization becomes an objective after the conceptual knowledge that underlies a procedure is developed. For most learners, mathematical concepts are first conceived of as process, algorithm, or action. Conceiving of a concept structurally requires that learners bridge a vast ontological divide wherein the concept solidifies into an object or structure; a process identified as reification (Sfard, 1994). It has been suggested that the development of mathematical concepts is a cycle of procedural-structural evolution (Sfard, 1991). Abstract mathematical concepts can be conceived of in two fundamentally different ways: structurally, as objects; or, operationally, as processes (Sfard, 1994). It appears that there exists a deep ontological gap between operational and structural conceptions. For example, mathematical concepts are initially perceived of as actions or processes (operationally). As further understandings are developed, the mathematical concept is conceived of as an object (structurally). In their consideration of the duality between process and concept, Gray and Tall (1994) hypothesize that there is a qualitatively different kind of mathematical thought displayed by more able thinkers compared to less able ones. The inability to conceive of mathematical concepts as objects rather than processes is a source of conflict and misunderstanding for many learners. Recognizing the interrelationship between conceptual and procedural knowledge, the authors describe proceptual thinking as a combination of conceptual and procedural thinking. They conjecture that procept is a process that produces a mathematical object and a symbol that represents either the process or the object. As proceptual thinking grows in conceptual richness, "procepts can be manipulated as simple symbols at a higher level or opened up to perform computations, to be decomposed and recomposed at will (Gray and Tall, 1994, p. 137).

Representations

In order to think or talk about mathematical ideas, they must be represented in some way. Internal representations are distinguished from external representations in that internal representations refer to the possible mental configurations of individuals (Goldin & Kaput, 1996). Because they cannot be directly observed, the presence of internal representations in the mind of a learner must be inferred from observations of what learners do or say. As opposed to learning as internalization, Lave & Wenger (1991), who advocate a sociocultural theory of teaching and learning mathematics, define learning as both a discursive activity and a form of participation in the activities of a community of practice. They reject the depiction of learning as assimilation of information from the external world to create a form of internal representation. Unlike internal representations, external representations are not mental processes, but physically embodied, observable configurations. An external representation is a configuration that corresponds to, is referentially associated with, stands for, symbolizes, interacts in a special manner with, or otherwise represents something else (Palmer, 1977). External representations include symbols, schema, diagrams, and models in both pencil-and-paper, concrete, and computer microworlds. However, as Dörfler notes, representations are not mathematical objects. He asserts, "none of the external representations can play the role of the intended mathematical object because no single one representation can be ascribed the totality of properties that the mathematical object is viewed to possess (Dörfler, 1996, p. 470). Representations are a powerful means for learners to communicate with and reflect upon their environment. They can also be a tool for dialogue both with oneself and with others (Lemut, du Boulay, & Dettori, 1993). Understanding mathematical concepts, as well as having the ability to solve mathematical problems, involves building and handling representations. Representations do not represent by themselves. Rather, they need to be interpreted and, to be interpreted, they need an interpreter. That is, an interaction or act of interpretation is involved in the relation between that which is representing and that which is represented (von Glasersfeld, 1987). Charles Peirce (1935), whose writings were a primary source of the theory of semiotics, asserted that, because we know the world through signs, which must be interpreted by other signs, all knowledge is indirect (Siegel & Carey, 1989). These conceptualizations of representations contrast with the notion that representations contain the knowledge that they are meant to convey and, as such, are transmittable from teacher to learner. According to Steffe and Wiegel (1996), many accounts of knowledge representation are based upon the assumption that concepts are things - mental objects - "out there" to be represented. They regard mathematical concepts as mental acts or operations; it is the operations that are represented. Further, representational elements are constructed as part of the construction of the concept. Basic representational elements are subject to modification and elaboration and may become associate with, or a constitutive part of, other representational elements. The use of external representations can be syntactically driven. That is, the learners may operate exclusively on the symbols themselves without referring to the internal representations involved. Learners who rely on a direct, syntactic approach to problem solving use a phrase-by-phrase translation of the problem into variables and equations (Chaiklin, 1989). For example, the learners who solved the problem: *"There are six times as many students as professors at this university"* by writing $6S = P$ assumed that the sequence of words mapped directly onto a corresponding sequence of literal symbols in an equation (Sims-Knight and Kaput, 1983). In contrast, in the case of learners who arrived at the correct solution, $6P = S$, the use of representations may be considered as semantically driven. Internal representations were called into play to assist students in linking the equation to the meaning of the problem (Herscovics, 1989). While the growth of mathematical knowledge has been described as a process of constructing internal representations of information and connecting these representations to form organized networks, understanding is the process of constructing and establishing relationships between these existing networks and new information (Hiebert & Carpenter, 1992). Meaning or understanding in mathematics comes from building or recognizing relationships either between representations or within representations.

Mathematical Understanding

The notion of what it means to understand mathematics is evolving from the behaviorists' notion of recall, retention, transfer, and mastery of a hierarchical sequence of skills and concepts to understanding as a constructive activity based upon problem posing, problem solving, investigation, and collaborative sense making (Davis, 1992). A dynamic view of the nature of mathematical understanding (Kieren, 1994) is based, in part, on Maturana and Varela's (1987) enactive view of cognition as an embodied, interactive process, coemergent with the environment in which a person acts. Traditionally, much of mathematics education has been dominated by behaviorist perspectives. A behaviorist pedagogy is characterized by a transmission model of learning. According to this paradigm, the teacher transmits a fixed body of factual knowledge and computational skills to learners. Learners who demonstrate the ability to efficiently manipulate symbols and recall facts are regarded as having mathematical understanding. As Goldin explains, "the exclusively behavioral characteristics of desirable learning outcomes leads educators to rely on the teaching of discrete, disconnected skills in mathematics, rather than on developing meaningful patterns, principles, and insights" (1990, p. 36). In contrast to behaviorist approaches to mathematics learning, as a theoretical perspective, constructivism explains mathematical learning as a matter of reorganizing activity, where activity is interpreted to include conceptual activity or thought, rather than a process of internalizing carefully packaged knowledge. A view of mathematics as socially constituted asserts that mathematical knowledge is appropriated through participation in socially defined communities of practice. Mastery of mathematical skill and knowledge requires learners to participate and become acculturated in the sociocultural practice of the mathematics classroom (Cobb, Yackel, & Wood, 1992). Similarly, the interactionist perspective argues that, "learning is characterized by the subjective reconstruction of societal means and models through negotiation of meaning in social interaction" (Bauersfeld, 1995). Mathematical understanding is actively constructed and reorganized as the learner relates what he or she already knows with a new experience or with new information arising from instruction (Davis, Maher, & Noddings, 1990). The construction of mathematical knowledge also requires that learners engage in the processes of reflection, recursion, and communication (Hiebert & Wearne, 1992; Kilpatrick, 1985). Reflection and recursion are conscious, internal mental manipulation processes that can be promoted and guided through instructional practices, including those that encourage and promote communication in the mathematics classroom. Vygotsky (1978) emphasized social interaction and interpersonal relations as well as the use of cultural tools such as mathematical symbols (Cobb, 1995). Learners appropriate tools for thinking by participating in socially defined systems of activity (Rogoff, 1990) or communities of practice (Lave, 1991). Cobb (1995) describes the process of learning to use a cultural tool, such as a mathematical notation system, as involving "the construction of an experiential mathematical reality that is consensual with those of others who can be said to know mathematics" (p. 122). The contradiction between individual construction (including the processes of reflection and recursion) and sociocultural processes (including communication) is alleviated when one considers that, according to both Mead (1934) and Vygotsky (1978), the tools that have developed to enable communication in society also enable the development of individual thinking. Acquiring the language and symbols of mathematics (tools) supports and enables the construction of mathematical knowledge. The vision of the mathematics classroom as a community of collaborative, reflective practice implies that new forms of instructional practice are required in order to support learning in this environment. If learning mathematics is indeed a social activity, learners must be provided with the language and the symbols (tools) that enable them to engage in reflection, recursion, and communication; and, ultimately, to construct mathematical understanding.

Teacher Knowledge

Like the relationship between conceptual and procedural knowledge, the relationship between teachers' knowledge and beliefs and their practice is dialectical (Cobb, Yackel, & Wood, 1991). While research has failed to support a direct relationship between teachers' knowledge of mathematics and student learning (McKnight, et al., 1987), teachers' knowledge of mathematics does have an impact on instruction in terms of its richness and conceptual level (Fennema & Franke, 1992). Many of the same misunderstandings and "weak conceptualizations" that have been identified in learners are also prevalent among teachers (Post, Harel, Behr, & Lesh, 1991). Ball (1990) asserts that, because there is a difference between recalling and performing and understanding, teachers do not develop a substantive understanding of mathematics in their own elementary and secondary educations. Consequently, many teachers are not prepared to effectively teach mathematics as envisioned by recent mathematics education reform efforts (NCTM, 2000). It appears that teachers' weak (or nonexistent) conceptual understanding of mathematics leads to an overemphasis on the teaching of procedural knowledge and the ineffective teaching of conceptual knowledge (Dunn, 2001). Although the task of investigating the construction of conceptual knowledge and the subsequent transformation of this knowledge into pedagogical content knowledge is extremely complex, it appears that there is a relationship between pedagogical content knowledge and instructional practices (Shulman, 1986). Shulman describes pedagogical content knowledge as the integration or synthesis of teachers' pedagogical knowledge (what they know about teaching) and their subject matter knowledge (what they know about what they teach). Shulman's work reveals that teachers with greater mathematical knowledge were more conceptual in their teaching while teachers with lower levels of knowledge were more rule-based. The study also revealed that teachers who possessed greater pedagogical content knowledge were able to provide learners with the multiple representations necessary to encourage conceptual learning. They had multiple ways of representing the subject to make it more comprehensible to learners, more extensive repertoires of explanations, examples, and demonstrations; and greater ability to draw on their knowledge during instruction. Cobb, Yackel, and Wood (1991) draw an analogy between teachers' construction of pedagogical knowledge and learners' construction of mathematical knowledge. The authors note that, because the direct teaching of formal mathematics leads to rote learning, learners should be encouraged to construct mathematical knowledge in informal settings. Analogously, they assert that the direct teaching of pedagogical content knowledge will lead to rote teaching. The authors suggest that teachers' should develop their pedagogical content knowledge by building on their informal ways of interpreting and drawing pedagogical interpretations of learners' mathematical activity. Subsequent studies support the finding that inexperienced teachers have incomplete or superficial levels of pedagogical content knowledge (Carpenter, et al., 1988; Feiman-Nemser & Parker, 1990). Moreover, novice teachers, even those with substantial content area knowledge (Grossman, et al., 1989), grapple with how to transform and represent concepts and ideas in ways that make sense to the specific learners they are teaching (Wilson, et al., 1987). Cobb, Yackel, and Wood (1991) suggest that teachers' beliefs and pedagogical content knowledge are profoundly influenced by their experiences of interacting with learners in the classroom. In an examination of the process of learning to teach mathematics for understanding, Eisenhart, et al. (1993) illustrate how several factors, including teachers' weak understanding of mathematics, lead to an overemphasis on the teaching of procedural knowledge and the ineffective teaching of conceptual knowledge. The researchers designed data collection instruments which enabled them to explore the process of learning to teach in numerous ways and from numerous perspectives. Data were analyzed using a framework which allowed the researchers to examine the teacher's knowledge and beliefs; her thinking and actions in the classroom; the sociocultural environment of the school; and, the sociocultural environment of the university. Analyses of classroom observations revealed that, while the preservice teacher possessed a strong procedural understanding of mathematics, because of a lack of conceptual understanding, her teaching was characterized by procedural rather than conceptual explanations. Moreover, analyses of interview data about the student teacher's knowledge and beliefs revealed that despite vocalizing a goal of teaching mathematics for understanding, the participant's weak conceptual underpinnings inhibited her teaching for conceptual knowledge. In a study of primary preservice teachers, Crawford (1992) noted that, although they had completed mathematics courses in which they had been successful in learning axioms and algorithms, few of the preservice teachers had any experience in using mathematics to solve novel mathematical problems or with talking about and explaining mathematical concepts. The researcher observed that this gap in their mathematical knowledge was only revealed as the preservice teachers attempted to think mathematically in a new setting.

Beliefs and Attitudes

Secondary mathematics teachers' knowledge, beliefs, and attitudes about learners and student learning have an impact upon the way in which these teachers interact with learners in their mathematics classrooms (Calderhead, 1984). In turn, learners' own attitudes, achievement, and persistence in mathematics are shaped, in part, by teachers' beliefs and attitudes (Stanic & Hart, 1995). Beliefs are distinguished from both knowledge and emotions. Thompson (1992) identifies two characteristics that distinguish beliefs from knowledge. The first is that beliefs can be held with varying degrees of conviction. A second feature of beliefs is that they are not consensual; the believer is aware that others may think differently. Beliefs are mainly cognitive (as opposed to physiological) in nature and are built up over a relatively long period of time. In contrast, emotions involve a physiological response, require little cognitive appraisal, and tend to change rapidly (McLeod, 1992). In contrast to emotions, which may appear quite suddenly, attitudes are affective responses that involve positive or negative feelings of moderate intensity that are relatively stable (McLeod, 1992). Examples of attitudes toward mathematics include liking algebra, being bored by geometry, being anxious about proofs, and disliking word problems. While attitudes may be difficult to separate from beliefs, attitudes play an important role in the mathematics classroom. For example, like beliefs, teachers' attitudes toward mathematics, such as boredom or anxiety, can be perceived by students and may have an influence on them (Benninga, et al., 1981). Moreover, national assessment data (Dossey, et al., 1986) indicate a positive correlation between learners' attitudes and their achievement in mathematics. According to McLeod (1992), attitudes toward mathematics seem to develop in two ways. Attitudes may result from automatizing a repeated emotional reaction to mathematics, such as a repeated negative emotional reaction experienced when simplifying algebraic expressions. Eventually, the emotional reaction will become automatic and more stable, developing into an attitude. Another source of attitudes is the assignment of an already existing attitude to a new but related task; for example, assigning a negative attitude toward geometric proofs to algebraic proofs.

Teachers' Beliefs and Attitudes

Ernest (1991) asserts that teachers' beliefs about mathematics, as well as the social context of the situation, influence mathematics teaching. Teaching practices reflect teachers' beliefs which, in turn, reflect their own experiences and background (Cabello & Burstein, 1995). Results from studies assessing teachers' beliefs indicate that the beliefs of teachers play an important role in influencing the interests, motivations, and aspirations of their learners. Because teachers act upon their expectations of learners, negative teacher beliefs or low expectations for their students influence classroom practices and may adversely affect student performance (Brophy, 1985). A classroom teacher's conception of mathematics has a powerful impact on the way in which mathematics is approached in the classroom (Cooney, 1985). There is a commonly-held perception of mathematics as a closed subject where there is only one right answer (and only one right way of arriving at it), and where divergent thinking is unnecessary (Lampert, 1990). Cooney suggests that the typical belief of the existence of a formal, external body of knowledge that must be transmitted to learners, rather than a dynamic, changing field in which learners can explore, experiment, and experience the processes of mathematics may make reform efforts, such as those envisioned by NCTM, slow and difficult to implement. Lampert concluded that mathematics is commonly associated with certainty, "knowing it, with being able to get the right answer, quickly." She found that within the school culture, doing mathematics means following the rules laid down by the teacher, knowing mathematics means remembering and applying the correct rule when the teacher asks a question, and mathematical truth is determined when the answer is ratified by the teacher. An investigation of middle school mathematics teachers' beliefs and their classroom teaching revealed evidence that teachers' beliefs, views, and preferences about mathematics and its teaching influence what they do in the classroom and play a significant role in the teachers' characteristic patterns of instructional behavior (Thompson, 1984). For example, Thompson noted that the teacher who believed that mathematics is an organized, logical system of symbols and procedures focused on establishing and maintaining order, presented content in a precise manner, and stressed the reasons and logic underlying mathematical rules and procedures. In contrast, Thompson observed that the teacher who believed that the domain of mathematics is continuously expanding and undergoing changes to accommodate new developments utilized a heuristic approach to presenting content and emphasized problem-solving activities that encouraged learners to conjecture and reason in order to experience the process of acquiring mathematical knowledge. Cooney's (1985) use of ethnographic methods to examine the beliefs that a first-year secondary mathematics teacher had about problem solving and how it should be taught confirmed that social context has an impact upon teachers' instructional decisions (Ernest, 1989). The teacher in Cooney's study experienced tension and conflict between his beliefs about teaching and learning and his classroom practice. Despite expressing a belief in the value of a problem-solving approach to teaching mathematics, when faced with learners' weak prerequisite knowledge and low levels of enthusiasm, the teacher abandoned problem-based lessons and, instead, resorted to drill and practice, skill-based lessons in order to maintain classroom control and cover subject matter (Cooney, 1985). While studies support the view that a teacher's behavior in the classroom is shaped by internal principles that are based upon beliefs about mathematics and mathematics teaching, it appears that the origins of teachers' beliefs are linked to experiences prior to their participation in mathematics education courses. Research has revealed that preservice primary teachers enter mathematics courses with strongly held attitudes and beliefs about how mathematics is learned and the role of the teacher that were derived from their own educational experiences (Crawford, 1992). These beliefs about learning mathematics - about what teachers and students do - shaped their perceptions of classroom events and their behavior as beginning teachers. For example, when asked to work in groups to develop a student-centered context for mathematical inquiry, nearly all of the preservice teachers experienced difficulties because of their belief that they needed to "teach the children something" before they could engage them in inquiry or investigations. They spent most of their instructional time telling the children the answer or showing them how to do it (Crawford, 1992).

Learners' Beliefs and Attitudes

Learners' beliefs about the nature of mathematics have been identified as the "invisible hand operating in mathematics instruction" (Borasi, 1990). Research suggests that social stereotypes and stages of intellectual development (Belenky, et al., 1985) as well as the way mathematics is presented in school (Borasi, 1990) play a role in shaping learners' beliefs about mathematics. Research also indicates that learners generally believe that mathematics is rule-based and mathematical problem solving consists of recalling and applying a series of memorized procedures to solve problems (Schoenfeld, 1985). Such beliefs may inhibit learners' ability to solve nonroutine problems. For example, a consequence of the commonly held belief that learning mathematics is a matter of practice alone is a resistance to reason about or understand mathematical ideas (Borasi, 1990). Learners who believe that underachievement in mathematics is due to low ability may attribute their failure to achieve to an uncontrollable cause (Weiner, 1986). When learners attribute failure to lack of ability, they exhibit a pattern of learned helplessness, demonstrate a low level of persistence, and avoid challenges (Diener & Dweck, 1978).

Mathematical Self-Concept

Research has determined a positive relationship between mathematical self-concept and achievement in mathematics (Marsh, 1986; Marsh, 1992). McLeod (1992) asserts, "The relationship of self-concept to achievement is consistently positive." Self-concept, a generalization of confidence in learning (Reyes, 1984), is a person's perception of self constructed through experience with the environment and influenced by both environmental reinforcements and the reinforcement of significant others (Shavelson, Hubner & Stanton, 1976). Mathematical self-concept can be considered to be a general belief about oneself that is stable and long-lasting. Moreover, it contains strong affective components. In their research on motivations to learn and persist in mathematics, Fennema and Peterson (1985) use the terms self-concept and confidence interchangeably. Learners develop a general self-concept, which can be analyzed into various components, including an academic self-concept (Shavelson & Bolus, 1982). Specific self-concepts, including one's academic self-concept, are combined to make up one's general self-concept (Byrne, 1984). Self-concept may be shaped by sociocultural elements, including school experiences. For example, identification as a low achiever or at-risk learner, placement in lower level mathematics classes, and low expectations of teachers may contribute to low self-concept. In an examination of the differences in self-concept among male and female elementary and secondary teachers, Relich (1996) found only minimal gender differences between male and female teachers' self-concepts, attitudes, and approaches to teaching mathematics. Data analysis did reveal that prior school experiences with mathematics and parental attitudes toward the subject had a strong influence on the self-concepts of the teachers in Relich's study. According to Relich, high self-concept teachers were more motivated, inventive, and creative in teaching mathematics. In contrast, the low self-concept teachers were more likely to complain about the lack of time and resources, and demonstrate limited creativity in their teaching. Relich concludes that teachers with low mathematical self-concept may provide inappropriate role models for their students. Given the relationship between teacher self-concept and student achievement (Midgley, Feldlaufer & Eccles, 1989; McLeod, 1992), and the pivotal role that teachers play in the mathematics classroom, it is possible that preservice teachers with low mathematics self-concepts may have a negative influence upon their own students' formulation of mathematics self-concept.

Mathematical Discourse

In mathematics, new knowledge is produced in a community of discourse (Davis and Hersh, 1987). Mathematical discourse includes the accepted rules of evidence and the methods of arguments that are used to validate mathematical ideas and procedures. The *Professional Standards for Teaching Mathematics* (NCTM, 1991) describes discourse as "the ways of representing, thinking, talking, agreeing, and disagreeing" (p. 36). Discourse in the mathematics classroom is impacted by educator's beliefs (Stanic & Hart, 1995) and includes the ways that ideas are represented, exchanged, and modified into more powerful and useful ideas (Lappan & Briars, 1995). Tharp and Gallimore (1988) refer to typical classroom talk as "recitation script." Recitation scripts consist of a teacher-initiated query, a student response, and an evaluation by the teacher. Forman (1992) notes that this form of classroom discourse privileges the teacher's evaluation of an answer and assumes that learning is a process of internalization rather than the negotiation of meaning. It provides no opportunity for learners to redefine the problem being presented or to offer alternative explanations, unless specifically requested to do so by the teacher. In contrast to recitation script, Tharp and Gallimore (1988) use the term "instructional conversation" to refer to classroom discourse that allows the co-construction of meaning between teacher and learners. Instructional conversations are modeled after everyday conversations in which more experienced members of a culture instruct the less experienced via talk. Learners are provided with opportunities to initiate topics, redirect discussion, debate issues, or provided their own explanations. Instructional conversations go beyond concept learning to include the mastery of specialized speech genres such as the mathematics register (Forman & Cazden, 1992). The linguist Halliday (1975) coined the term register to describe "a set of meanings that is appropriate to a particular function of language, together with the words and structures which express these meanings" (Halliday, 1975, p. 65). Cazden (1986) describes a register as a sociolinguistic construct. The mathematics register extends the use of natural language by giving words new meanings that are particular to a role or function. According to Pimm (1987), the most striking characteristics of the mathematics register are "the number of terms that have been borrowed from more everyday English. Examples of such words include: face, degree, relation, power, radical, complete, integrate, legs, product " (Pimm, 1987, pp. 76-77). In traditional classrooms, where recitation scripts restrict participation, learners are asked to absorb the mathematics register by reading text and listening to the teacher (Forman & Cazden, 1992). Learning experiences that encourage learners to explain and construct different solutions to problems or compare multiple representations provide learners with the opportunity to employ and become fluent in the mathematics register. Lampert (1986) advocates that learners be actively involved in constructing their own solutions by applying their understanding of other mathematical principles to a variety of problems. In her examination of classroom discourse, Lampert noted that learners were given activities in which the mathematical principles were not taught directly nor formally named. Instead, language was utilized to develop an understanding of the underlying principles within the context of algorithmic instruction. Instruction was organized so that learners could actively engage in doing and talking about mathematics, including explaining and justifying their thinking to both their teacher and their peers. Lampert demonstrates that when learners came to solve the algorithms in the conventional way, they were able to explain their actions and justify their procedures by talking meaningfully about the mathematics. This suggests that classroom discussions facilitated learners viewing mathematics as more than a set of procedures for finding answers. Hiebert and Wearne's (1993) analysis of classroom discourse demonstrates that the verbal interactions between the teacher and learners and amongst the learners themselves play a significant role in learning mathematics. Their comparative study indicates that, in the classrooms where the most significant gains were made, the teachers were not the primary source of authority but, instead, play a role in interpreting learners' explanations and facilitating social interaction and the negotiation of meaning. Hiebert and Wearne's study implies that mathematics educators face two major challenges in designing a classroom environment that facilitates discourse. First, learners must be free to express their thoughts without fear of reprisal or sanctions. Additionally, learners must be encouraged to listen and to respect the ideas of their peers. Borasi (1996) identifies three levels of mathematical discourse at which inquiry occurs. The levels of discourse are defined by whether one is engaged in: (1) performing a specific mathematical task, solving a problem, attempting to provide a result; (2) understanding some technical mathematical content such as a concept, rule, or topic; or, (3) understanding the nature of mathematics which may involve understanding the notion of proof, definition, or algorithm, appreciating what characterizes mathematical thinking and mathematics as a discipline, and becoming aware of heuristics as well as of their domain of applications and limitations. Borasi notes that, while these three levels of mathematical discourse are complementary to one another, they may contribute to the achievement of different goals, depending upon whether the stance of learning is remediation, discovery, or inquiry. For example, when the stance of learning assumed in a lesson is inquiry, discourse at the level of understanding some technical math content can motivate questions that may lead to new perspectives and insights on a mathematical concept, rule, or topic not addressed in the original lesson plan. The differences in the types and amount of discourse in mathematics classrooms noted in several studies (Lampert, 1990; Cobb, et al., 1992; Ball 1993) suggest that teachers must have a conceptual understanding of mathematical concepts in order to orchestrate mathematical discourse that provokes and challenges learners' thinking and asks learners to clarify and justify their ideas. The multiple research perspectives discussed in this chapter provide insights into and a framework for examining issues related to students on the margin and mathematical achievement. This book draws inspiration from these research perspectives and seeks to draw together the findings from these works to inform practice and overcome achievement disparities.

For Exploration and Reflection

1. Visit a university library and conduct a literature search for research studies on marginalized students and mathematics teaching and learning. Identify a theme and write a thematic literature review to document and support this theme. Identify an area related to students on the margin and mathematical success for future research and provide a rationale for why a study of that area is important.
2. Interview a mathematics teacher and ask him/her to describe his/her perceptions of marginalized/at-risk students and mathematics teaching and learning. What problems and concerns do they identify? What does the teacher believe is the most difficult aspect of teaching mathematics to marginalized students? What efforts have they made to assist these students?
3. Interview a mathematics teacher about mathematics teaching and ask about his/her knowledge of students on the margin, his/her instructional practices, and self-assessment of these practices. Observe the teacher's classroom during a mathematics lesson and note any differences between the teacher's espoused and enacted behavior.
4. Visit several mathematics classes and observe any strategies employed by the classroom teachers to build and maintain a learning environment that encourages mathematical success by all learners. What strategies did the teachers you observed employ to motivate and keep students

motivated to learn mathematics? How did the teachers encourage students to explain their reasoning and engage in mathematical discussions?

Chapter Three

Methods of Inquiry and Analysis

The methodological underpinning of the study discussed here is derived largely from orientations to research that draw attention to the importance of detailed qualitative fieldwork and the observation and analysis of participants in contexts (Goetz & LeCompte, 1984). Seeking to understand what participants do and what they know that makes their actions understandable (Spindler, 1982), the researcher, a graduate student and former high school mathematics teacher, spent a semester as a participant-observer, making long-term observations in an alternative high school.

Participants and Context

The participants were selected using purposeful sampling strategies (Patton, 1990) and included five preservice secondary mathematics teachers and five students who attended an alternative high school. The preservice teachers were enrolled in a field-based mathematics methods course during the final semester of their teacher preparation program. The learning environment for the course is based upon the NCTM *Standards* (2000) and models inquiry-based, student-centered, collaborative learning. As part of the field experience component of the course, preservice teachers participate in a semester-long pre-student teaching experience. Because they were concurrently enrolled in the mathematics methods course, the preservice teachers had access to a variety of reform-oriented methodologies and materials that they could incorporate into their mathematics lessons during their field experiences. The learners had been identified by their teachers as marginalized, disconnected, or labeled "at-risk" and, at the suggestion of their counselors, had voluntarily elected to attend an alternative high school where they were enrolled in the semester-long integrated mathematics course taught by the preservice teachers.

Data Collection

Consistent with the methodology of the study, qualitative data were triangulated via multiple sources of evidence, including: (1) observations, videotaping, and field notes of the preservice teachers' mathematics lessons; (2) three semi-structured, open-ended interviews (Spradley, 1979; Novak & Gowin, 1984) with each preservice teacher focused on beliefs, perceptions, mathematics content, and instructional practices; (3) four semi-structured, open-ended interviews with each learner focused on beliefs, self-concept, and mathematics content; and (4) classroom artifacts including the preservice teachers' reflective journals, lesson plans, and copies of student work. Qualitative data were analyzed utilizing a double coding procedure (Miles & Huberman, 1994) and major themes were developed using thematic analytic strategies (Spradley, 1979). Qualitative data were supplemented and supported by quantitative data, including included pre- and post-study tests, rubrics, and questionnaires. Mathematics content knowledge was measured using pre- and post-study mathematics tests designed to access conceptual knowledge of mathematics concepts. Quantifiable data on the participants' beliefs and self-concept was elicited using pre- and post-study questionnaires designed to measure mathematics content knowledge, beliefs, learning patterns, and instructional practices. For example, information on self-concept was obtained from questions that probed the participants' methods of verification and attributions of success and failure. A pre- and post-study rubric with a scale from one to four provided quantifiable data on the preservice teachers' instructional practices. Standards-based categories on this rubric included content, communication, reasoning, problem solving, connections, and representation.

Data Analysis

In order to ensure interpretive validity, the researcher employed five strategies that provide a framework for interpretive validity in qualitative research (Miles and Huberman, 1984): (1) continuous revision of data collection methods; (2) keeping data contextualized; (3) various types of triangulation, including member checks; (4) using several analysts to check one another's interpretations; and, (5) preservation of the evidential chain. These strategies were employed as a way of verifying the representativeness of observed events and cross-checking interpretations of the meanings of those events. Quantitative data were analyzed by computing statistical means and standards deviations. Results obtained through quantitative analyses supported the findings from the qualitative data analyses.

Instructional Practices

The preservice teachers' instructional practices were examined within the context of a four-tiered scheme. Lessons were characterized as: (1) teaching for understanding, which is defined as conceptually-focused, active and student-centered; (2) conceptually-focused teaching, which is defined as teaching that is focused on concepts, structure, and connections but does not involve a significant degree of student involvement; (3) procedurally-focused with student involvement, which is defined as teaching that is focused on algorithms, procedures, and sequences and is student-centered; and, (4) procedurally-focused without student involvement. In addition lesson plans were scrutinized for instances of student-centered activities that focused on concepts, structure, and connections. In addition, the preservice teachers' lesson plans were analyzed using the same criteria as used for the analysis of instruction in regard to a conceptual and procedural focus as well as specific instructional practices. Because understanding is a constructive activity based upon problem posing, problem solving, investigation, and collaborative sense making (Davis, 1992), when determining if the preservice teachers were teaching for understanding, particular attention was paid to the use of activities that: (1) provided the learners with novel problems placed within familiar contexts; (2) encouraged learners to articulate their conceptual understandings; and, (3) allowed learners ample time to reflect on the nature of mathematical concepts (Hatano, 1988). Instructional practices that focused on the structure and connections between mathematical concepts and emphasized active reflection, discussion, or critical thinking were interpreted as promoting understanding and the construction of conceptual knowledge. In addition, the researcher examined the data to find instances of student-centered activities in which the learners were required to: (1) pose problems as part of their approach to solving problems rather than to merely produce a solution or response to a problem; (2) respond to open-ended versus closed ended questions; and, (3) find solutions to problems posed through examples drawn from real life. With regard to teaching that was procedurally-focused, the researcher looked for instances of activities that emphasized rote imitation and the mastery of computational skills. The analysis focused on activities that required learners to recognize symbols, follow rules, carry out sequential actions, and master procedures and algorithms.

Mathematical Development

The analysis of the learners' mathematical development with regard to skills and understandings focused on performance and comments that illustrated understandings of three mathematical topics that were central to the course instruction: (1) spatial sense and geometry; (2) number sense, including proportional reasoning, estimation, fractions, decimals, and percents; and, (3) functions and graphs, including variable, covariance, correspondence, and slope. Additional mathematical topics, such as literal expressions and graphing emerged but were not analyzed fully. In addition to interview data, the learners' participation in classroom interactions associated with these topics were used as indicators of learner understanding. These data also provided additional information on the learners' general problem solving behaviors. In addition to these analyses, computationally-focused tasks were examined for: (1) accuracy; (2) problem solving strategies, including a reliance on algorithms; and, (3) symbol system use in problem solving. In order to obtain data on the learners' spatial sense and their knowledge of geometry, the researcher asked learners to discuss relationships between geometric figures and make observations of similarities and differences. Learners were shown a diagram of a rectangle and asked to determine its area. They were also shown a parallelogram and asked to find its area. The learners were then asked to discuss the similarities and differences between the two figures, including observations with regard to the figures' areas. The researcher and learners revisited these polygons during a later interview session. Once again the learners were asked to describe similarities, differences, and note the characteristics that made them similar or different. The learners were also asked to consider the interior angle measures of the two figures. In order to acquire additional data on the learners' sense of the processes of finding area, on the written assessment the learners were asked to find the length of a diagonal of a rectangle when given the length and width. The learners were also required to find the area of a geometric shape composed of a semicircle, circle, triangle, and rectangle. One measure of the learners' spatial sense was obtained from the written assessment. Specifically, the learners were shown a diagram of a stack of cubes and asked to determine how many faces were showing. In addition to solving the problem, the learners were asked to articulate their problem solving strategies.

In an effort to assess the learners' ability to transfer their understandings of geometry to novel situations, the learners were asked to describe strategies they would utilize to find the area of their feet. They were also shown diagrams of a parallelogram and a trapezoid and asked to determine how they would find the areas of those figures. In order to gather information on the learners' number sense, the researcher gave the learners several tasks and observed whether they were performed in a non-computational manner. The learners were required to perform estimation tasks and to place fractions and decimals in order from smallest to largest. Another measure of the learners' number sense was obtained from a task that displayed a series of equilateral triangles stacked to build larger triangles. The learners were asked to identify the pattern and determine how many triangles it would take to build a structure ten triangles high. Data on the learners' understandings of functions and graphs were obtained from several sources. The learners were shown two linear equations and asked to determine which line would be steeper and explain their reasoning processes. The learners were also presented with a story context and asked to determine if it was a function. They were then asked to solve the problem and determine if slope was represented in the equation. The learners were then asked to describe slope and to give a real life example of it. During a subsequent interview, the learners were once again presented with a story context related to the concept of function and discuss their problem solving strategies. All of these data were analyzed for specific understandings of the general concept of function. These include the notion of variable, correspondence, covariance, translation, and relationships between symbol systems such as graphs and equations. In addition, story problem solving and discussion were analyzed for evidence of understanding of specific aspects of linear functions, including slope and intercept. The researcher also attempted to determine the learners' ability to transfer their understandings of functions and graphs to novel situations. The learners were presented with a variety of functions in various forms and asked to determine if they were the same or different, or linear or nonlinear, functions.

For Exploration and Reflection

1. Identify one or two fundamental issues related to mathematics education and students on the margin. State what methods you would use to research the issue. What data would you collect and how would you analyze your data? Explain why you would use that approach and justify your reasons for selecting these methods.
2. Interview a student who has been identified as marginalized or disconnected and ask him/her about his/her attitudes toward and perceptions of mathematics. Examples of some questions that you might ask are: "What comes to mind when you think of mathematics?" "How useful is mathematics in life?" "How do you feel about mathematics?" "What does mathematics mean to you?" "What do you feel like when you are doing mathematics?" "Do you know anyone who is good in mathematics?" "How do you know" "Do you think being good in mathematic is something that you are born with a talent for or can it be learned?" Reflect on the student's responses. What did you learn about this student's attitudes toward and perceptions of mathematics? What strategies would you implement to assist the student?
3. Interview a student identified as "at risk" of being marginalized in the mathematics classroom and ask him/her to solve a problem from a reform-oriented mathematics textbook. As the student solves the problem, ask him/her to share their problem solving strategies. Examples of some questions that you might ask are: "Do you have an idea of what might be a reasonable answer?" "What did you do first? "Why did you decide to do that?" "Can you solve it another way?" "Do you observe any patterns?" "What problems have you seen that are similar to this?" "What makes this problem difficult or easy for you?" "If your friend were struggling with this problem, how would you explain it?" "How do you know your answer is correct?" Describe the student's strategies and behaviors as they solved the problem. What did this activity tell you about the student's problem solving strategies and mathematical understandings that you couldn't learn from a test? Did

any of the student's strategies or behaviors surprise you? What activities or tasks might you plan to assist this student?

Chapter Four

The Preservice Teachers

This chapter is divided into seven sections. The first two sections will introduce and provide background information on the preservice teachers and the learners. Because teachers' beliefs about mathematics influence mathematics teaching (Ernest, 1991), section three will explore the beliefs held by this particular group of preservice teachers. Similarly, because the occurrence of lowered expectations for students on the margin suggests negative implications for the effective education of these learners (Secada, 1992), section four will examine the preservice teachers' perceptions of marginalized and disconnected learners who have been labeled "at-risk." Section five will explore the instructional practices employed by the preservice teachers during their pre-student teaching experience in the alternative high school. This section will also examine the preservice teachers' efforts to teach for understanding and their use of (1) technology; (2) collaborative learning groups; (3) and, student comments to facilitate discussion. In order to examine the impact of the pre-student teaching experience, Section six will present a profile of each of the preservice teachers. Three of these profiles are in the form of in-depth case studies and are designed to provide some insights into the totality of an individual preservice teacher's experience. Because teaching practices reflect teachers' beliefs which, in turn, reflect the teachers' own experiences and backgrounds (Cabello and Burstein, 1995), the relationships between the beliefs and instructional practices of Brad, Kevin, and Toni are investigated. In addition, the changes that occurred in the beliefs and instructional practices of these preservice teachers over time and across mathematical topics are examined. The final section of this chapter will summarize the effect of the teaching experience on the preservice teachers.

The Preservice Teachers' Backgrounds

Brad

Brad was raised on a farm in a remote and sparsely populated region of the state. There were 100 students in his high school and 20 in his graduating class. Both of Brad's parents, as well as his older brother were alumni of the university. Brad, his brother, and his father are all members of the same fraternity. The family continues to maintain close ties to the university, attending every home football game and parents' weekends. Although he earned a degree in agricultural engineering, Brad's father is a third generation wheat farmer. At the time of the study, Brad was a 21 year old college senior with no prior teaching experience. Brad graduated from high school with a 4.0 grade point average. He indicated that, at the university level, he had either taken or "worked with" differential calculus, integral calculus, advanced calculus, differential equations, linear algebra, probability, history of mathematics, combinatorics, number theory, Euclidean geometry, and projective geometry. Brad maintained that he had positive experiences with mathematics at the university although, "some of them were more interesting to me than others." Until the spring semester of his junior year, Brad maintained close to a 4.0 average. During the spring semester of his junior year, the "C" Brad earned in a differential equations class brought his grade point average down to 3.57. When asked about his decision to teach, Brad explained, "I always respected my teachers and I thought they had a tough job." Brad was inspired to teach by his high school mathematics teacher, who in addition to teaching all of the mathematics classes, was also a wrestling coach. Notwithstanding his plans to become a mathematics teacher, farming and the land remain an important aspect of Brad's life. During the course of the study, Brad often talked of life on his family's farm, indicating that he enjoyed working with vehicles and other mechanical devices. On several occasions, he suggested that he might ultimately return to farming after pursuing a teaching career.

Kevin

Kevin grew up in an affluent bedroom community outside of the state's largest city. His father was a heavy equipment operator for a power company and his mother was an elementary school teacher. Several times during the study, Kevin spoke of his life-long desire to become a doctor. He cited his interest in medicine as a reason for taking mathematics and science classes. However, after working in a hospital throughout high school, Kevin concluded that, as a doctor, he would not be able to have the same level of personal contact with people that he would as a teacher. Kevin cited his high school English teacher as a factor in his decision to become a teacher:

> I had a English teacher in high school that really, just through her openness and, she was just a very encouraging person and through that I think she

> really, she really pushed me into doing something where I feel I could do the same type of thing, not necessarily in teaching but working with people in a way that could be effective and, and, it ended up that I'm going into the teaching field so, she was an influence.

It appears that Kevin's decision to teach mathematics and science was shaped, in part, by his belief that all learners had the potential to achieve in mathematics:

> One of the reasons I'm going into the science and math fields, in education, I mentioned this area is because, I mentioned this earlier, I get very frustrated, especially by females that don't think they can do math and science and the only reason I think that is because they haven't done it in the past - I think everybody has a potential to succeed, there's, there's those that are, that it'll come very easy to and there's those that might take a little longer but I believe everybody has the ability to succeed.

Because he was pursuing a degree in physical science, Kevin had taken fewer mathematics courses than the other preservice teachers, "I've had differential calculus, integral calculus, Euclidean geometry, linear algebra; currently I'm taking probability and statistics." In addition to the fifteen credits of mathematics courses, Kevin had taken a variety of science courses which included eighteen credits of chemistry as well as courses in physics, biology, and the life sciences. At the time of the study, Kevin had a 3.5 grade point average. While he indicated that his experiences with mathematics courses at the university level had been positive, he revealed that his failure to take precalculus or calculus in high school made him less prepared for university-level calculus:

> Well, I think because I came in and the math placement test placed me relatively high, they put me in calculus, which was fine with me because that's what I needed to take to get where I wanted to go the only problem was, was that I would guess 90% of the people there, in that class, well most people were engineers, they had had calculus in high school and this was kind of a review and, and those people, those people I met that didn't have calculus, I didn't meet one person who at least didn't have precalculus and I didn't even have that, so my choice by not taking precalculus was a bad one, it wasn't so much my high school, it was my choice and for that reason, I was not extremely prepared but I coped and I think I got a B+ so it's nothing to cry about, I had to work at it.

Toni

At the time of the study, Toni was forty-four years old and the single parent of two teenage children. Prior to entering the teacher preparation program, Toni had a variety of careers, including working as a reporter and a graphic artist as well as a stint in the Coast Guard. She attended a private, parochial school for girls but dropped out of the ninth grade at age fifteen. Toni cited "traumatic" experiences with mathematics instructors as factors in her decision to drop out of school:

> I was called to task in front of the whole class because the teacher did not understand the way I did the problems, I was asked to do it according to the new math and I had grown up doing it according to the traditional way, so what, when I went up to the board to do a problem, it was sort of a combination of old math, new math, but I never showed everything on the board and I was accused of cheating - the teacher said "you had to have cheated" - her exact words were "you had to have cheated to get this answer because there is no way you could have gotten that answer doing what you just did on the board" - and I was like, that was it for me - turned me totally off - and as a matter of fact, that was about the time I started to fade off from school.

Toni eventually earned her GED and then went on to attend several community colleges. In addition to working toward a teaching certification, Toni was also obtaining a degree in general studies with emphases in mathematics and Spanish. At the time this study commenced, Toni had a 3.6 grade point average. She indicated that she had taken university-level mathematics courses that covered differential calculus, integral calculus, differential equations, advanced calculus, Euclidean geometry, projective geometry, number theory, linear algebra, topology, probability, and, history of mathematics. Toni revealed that her experiences with mathematics courses at the university level had been "mixed":

> Overall I'd say it's been pretty good because I've managed to find some strong women that are understanding and are willing to listen to individual priorities and ways of doing things. Frustrating, in more outside the math, in coping with the system because of my problems with testing and trying to get help with that - and forget about getting A's - it's a shame that so much of the grade is contingent on testing because that's not what math is all about, math is more that you understand what you're doing and the tests don't always measure that.

Toni's decision to become a mathematics teacher was strongly influenced by her own negative experiences with school mathematics as well as by her conviction that women are acutely underrepresented in mathematics:

> Oh, my decision to teach math had to do with my experiences with math teachers and the fact that I think both boys and girls need to see two things, they need to see people teaching math that like math and women liking math and teaching it - I just think those things are important for both boys and girls and that's also why I chose the middle school level because I have done some research on my own about girls dropping out and that's the age where boys are coming in and thinking girls are secondary citizens as well, not as smart and those two things made me decide that's why I wanted to teach math.

Melissa

Melissa was raised in a large mill town on the west coast of the state. She earned an associates degree from a community college and a Bachelor of Science degree in mathematics, with a minor in economics from a small state college. After graduation, Melissa then entered graduate school at a university located in a neighboring state and intended to earn a masters degree in mathematics. She was awarded a teaching assistantship and taught precalculus. It was there that Melissa discovered that she enjoyed teaching. Unfortunately, her father died during her first semester and she was unable to continue her studies. Melissa returned to her family and accepted a position as an actuary at a large insurance company. However, Melissa explained that she missed the contact with people that teaching had provided:

> Well, originally I wanted to go into, like, to be a statistician or to be an actuary and I kind of got into it and I worked for an insurance company and really, really explored the actuary end of it and I'm just, I'm a people person and their work is a lot of independent work, they do a lot of work on their own, and I, I really enjoy math, and I like to help, I've tutored a lot, and I like to help other people, and when I taught at Harbor State I thought it was fun.

Soon after she had made her decision to pursue a teaching career, Melissa returned to college to earn a teaching credential in mathematics. At the time of the study, Melissa was twenty-six years old and in her last semester of a two semester certification program. Melissa, with her previously earned bachelor's degree in mathematics, had a stronger mathematics background than the other four preservice teachers. Her mathematical knowledge and ability were evident in her discussions about mathematical content areas. She indicated that she had taken the following mathematics courses at the university level:

> Differential calculus, integral calculus, differential equations, advanced calculus, Euclidean geometry, number theory, projective geometry, linear algebra, probability, history of math, Fourier series, which is just an extension of differential equations, I've taken a lot of statistics, probably like 2 years of just like, what else have I taken? I can't remember the name of the class, it was kind of like combined with computers, I can't remember the name of it.

Melissa indicated that she had positive experiences with these mathematics classes. During her first semester at Inland State, she maintained a 4.0 grade point average. While grades for Melissa's second and final semester were not finalized at the conclusion of the study, Melissa said, "I'm getting all A's." Although Melissa cited that her desire to work with people was a reason for her entering the teaching profession, subsequent conversations revealed that Melissa's mother had also played an important role in her decision to become a teacher: "Oh, my mom would just be so disappointed if I didn't become a teacher! She just thinks that teaching is such an important job, she is so impressed with it, with teachers."

Mark

Mark was born and raised in a suburb of the state's largest city. His father, now retired, was a newspaper editor for twenty-four years. His mother was a counselor for community college students and for women from battered homes. Mark frequently discussed the social work his older sister was doing in Central America. Throughout the study, he expressed his tremendous admiration for her and the impact she was having upon people's lives. Because of personal problems, including the death of a close friend, Mark had dropped out of college. After two years of teaching mathematics as a teacher's aide at an alternative school for learners with severe behavior disorders, Mark returned to college to complete a bachelor's degree in mathematics and earn a teaching certification: "First, I don't know whether becoming a math teacher is exactly what I had in mind, teaching, getting my endorsement, the teaching was my first priority - and I like math, I'm, I'm a credit short of getting a math degree so, might as well..." Mark indicated he had taken basic differential equations, three calculus courses, two linear algebra courses, two probability courses, and optimization modeling. The circumstances that resulted in Mark's dropping out of school for several years inhibited his achievement in these courses and he struggled to maintain a 2.5 grade point average. While the following comment provides some clarification as to Mark's academic difficulties, it also sheds some light on his personal desire to understand mathematics "in and out" and "backwards and forwards":

> My knowledge of these is, is really very, some classes, my attitude toward school, till, till I came back was like either I liked the proof and I enjoyed the material and I studied really hard just because I wanted to or if I didn't I would really skate, and so, (laughs), some material like geom - like calculus I know in and out and probability I know pretty well from what I've had but, linear algebra, that's why I'm retaking it, because I really never... I got, I understand the basics but I don't have a really thorough backwards and forwards understanding of it, just that I'd like to know that.

At the time of the study Mark was a 27 year old senior and in his first semester back at Inland State. Of the five preservice teachers, Mark was the most open and candid when discussing his feelings and experiences in the classroom. He often came to the researcher for feedback, critique and suggestions for ways to improve his teaching. He vacillated between embarking upon a career as a special education mathematics teacher, which he viewed as service to humanity, and using his mathematical skills to pursue a corporate career. Toward the end of the semester, Mark discussed his decision to apply to graduate school in order to pursue a Master of Science degree in mathematics. At the conclusion of the study, it appeared that his plans to embark upon a teaching career were on hold while he sought admittance to graduate school.

The Learners' Backgrounds

Johnny

Johnny was a nineteen year old senior who at a local fast food franchise. His job there was often a topic of conversation among both the learners and the preservice teachers. Johnny's mother worked in a clerical position at the university and his father was employed by a local automobile dealership. Johnny was a rather quiet student with a dry sense of humor. He was close to all three of the male preservice teachers. However, he had developed a special bond with Toni, who made frequent visits to the fast food outlet and conversed in Spanish with Johnny while he worked the night shift.

Rianne

Rianne was a seventeen year old junior. Her mother, who did not finish high school, was on medical leave from a custodial position at the university. After a series of negative experiences in the conventional high school, Rianne elected to enroll in the alternative high school; her third high school in three years. Her counselor revealed Rianne came from a "disturbed and troubled" home but was afraid to disclose any of these problems because she feared the school would notify the authorities.

Mollie

Mollie was raised in three different states. She was fifteen years old and lived with her father who had returned to school as a nontraditional student. Her mother, who lived in a neighboring state, was an English teacher at an alternative high school. Mollie's outgoing personality, coupled with the fact that she was a quick study, made her a favorite of the preservice teachers. However, like the other learners, Mollie's background was not without problems. Her parents and siblings were scattered across four states. At the conclusion of the study Mollie explained that she and a younger brother would spend twenty-one hours on a bus in order to arrive at her mother's home in time for Christmas.

Jamie

Jamie was a fifteen year old sophomore who continued to maintained close ties to her Native American community in a neighboring state. Jamie's parents both held professional positions at the university. Jamie had spent a year and a half in jail for a variety of offenses, including shoplifting and substance abuse and was out on parole. On two occasions during the course of the study, she violated her parole and was returned to a juvenile detention facility. She continued to battle substance abuse throughout the semester. Despite her personal difficulties,

Jamie was an extremely likable and intelligent student who won the affection and concern of the preservice teachers.

Wendi

Wendi was sixteen years old and had lived in ten different geographic locations. Wendi, her siblings, and single mother continued to relocate from one apartment complex to another. Her father, who was living in Florida, had earned a GED. Wendi openly discussed her personal and family problems in class. For example, during one class session, Wendi confessed that her father, who was thirty-eight, "was in bad shape and won't make it to forty" because of his addiction to drugs and alcohol. Wendi herself was plagued by a variety of personal, family, and behavioral problems which appeared to have a tremendous effect on her attendance as well as her ability to stay focused and function in class. Shortly before Thanksgiving, recognizing that she was "out of control," Wendi voluntarily enrolled in a drug and alcohol rehabilitation program and was unable to complete the mathematics course.

Preservice Teachers' Beliefs

Research supports the view that teachers' instructional practices are shaped by their beliefs about mathematics and mathematics teaching (Thompson, 1984; Cooney, 1985). Therefore, the purpose of this section is to examine the beliefs held by preservice teachers. As the analyses of these data unfolded, categories of beliefs about mathematics, mathematics teaching, and mathematics learning were further refined and clustered into the following groups: (1) mathematics as a tool consisting of a body of procedures vs. mathematics as a system of conceptual knowledge; (2) success through talent vs. success through hard work; (3) teaching through transmission vs. teaching through discovery activities; and, (4) one solution method vs. multiple solution methods.

Mathematics as a Tool Consisting of a Body of Procedures vs. Mathematics as a System of Conceptual Knowledge

With the exception of Kevin, the preservice teachers expressed both intrinsic and extrinsic reasons for learning mathematics. On questionnaire items that probed Kevin's beliefs about mathematics, he was the only preservice teacher who cited exclusively intrinsic reasons for learning mathematics. Kevin's belief in the intellectual importance of mathematics remained constant over the course of the semester and was evident in observations of his efforts to utilize investigative, open-ended teaching strategies. However, at the conclusion of the semester, Kevin revealed that teaching at Transcend challenged him to examine his own understandings:

> It's interesting teaching math because, I have to go back and re-learn it, I'll know the stuff and I can do the stuff but to teach math, you need to understand all the concepts, it takes up, it takes a lot to rethink, ok, why is

> this, and, what is this, so I learned a lot of, and I know I will re-learn a lot of stuff in math when I'm teaching it, just, stuff that I know, the concept of, but then I have to, to really explain, you have to know all the concepts so, you almost have to re-learn math that way so you can you explain it, effectively.

Mark believed that mathematics served two roles. The first role was to prepare learners for college and technical fields. The second role was a bit more complex:

> It [mathematics] uses a certain part of your thought that I think a lot of people don't develop very well, like logical thinking, it's really useful to have that in all life experiences, being able to rationalize things out as opposed to, I don't know, as opposed, as opposed to always following your emotions, to have intellectual thought, rational, intellectual encounter with that.

These two roles are consistent with Mark's questionnaire responses in which he cited both intrinsic and extrinsic reasons for learning mathematics. Observation and interview data confirm that the intellectual challenges offered by mathematics were more important to him than mathematics' utilitarian purposes. Although Toni also cited both intrinsic and extrinsic reasons for learning mathematics, she also made reference to the fact that she believed mathematics was vital to the development of critical thinking and reasoning skills. Observations of Toni and Mark's efforts to teach conceptually revealed that their beliefs were connected to their instructional practices. At the conclusion of the semester, these beliefs appeared to be unchanged. Both Brad and Melissa expressed utilitarian views of mathematics which remained fairly consistent throughout the semester. Engaging in the pre-student teaching experience appears to have done little to transform Melissa's beliefs about mathematics. For example, at the conclusion of the semester, Melissa once again identified the most important reason for teaching mathematics as "there's a lot of decisions in your life that involve math." However, at the end of the semester, Brad explained how participating in the teaching experience had changed his beliefs about mathematics, "The fact that math isn't always one right answer, there's a lot of, exploration that can go on instead of just hammer out problems with one right answer."

Success through Talent vs. Success through Hard Work

The preservice teachers held diverse beliefs about the role of talent and hard work in mathematics learning. For example, while Toni was the only preservice teacher who believed that success in mathematics was due exclusively to hard work, both Toni and Kevin expressed a solid belief that it was not possible to have an inherent talent for mathematics. During subsequent discussions, Kevin stated that success in mathematics was due to a combination of talent and hard work. Kevin qualified this statement by explaining that he was referring to high school mathematics which he described as relatively "concrete." According to Kevin, hard work would ensure success in school mathematics regardless of whether one possessed an inherent talent for the subject. However, he noted that once the mathematics became more abstract, "like what they do here" [the university's mathematics department], talent played a greater role than hard work. While Mark noted that success in mathematics was a result of both talent and hard work, he believed that it was possible to have an inherent talent for mathematics. He explained that those without a natural talent could still learn mathematics, but "it takes a lot more work." For those with a natural talent in mathematics, "You can always learn it, it's something you can learn without as much work." Although Melissa and Brad shared Mark's belief that success in mathematics was a result of both talent and hard work, they indicated that they believed it was possible to have an inherent talent for mathematics. Moreover, all of the preservice teachers attributed their own success in mathematics to both hard work and the fact that they were "good at math."

Teaching through Transmission vs. Teaching through Discovery Activities

The preservice teachers held a variety of beliefs related to transmission and discovery in mathematics learning. The majority of preservice teachers revealed that they learned mathematics best by employing a variety of strategies, including memorizing a series of steps, watching and copying others and working through the problems on their own. However, these beliefs related to their instructional practices in dissimilar ways. For example, although Melissa indicated that she learned mathematics best by utilizing a variety of methods, her procedurally-focused teaching appears to have encouraged learning mathematics by memorizing a series of steps. While Brad felt most strongly that he learned best by going back and working through the problems on his own, his teaching often led to the students following along with a specific procedure which he demonstrated. However, during the final interview session, Brad indicated that the experience enabled him to develop less of a single strategy, transmission belief and more of a multi-method, student-centered notion of teaching, "Just as more of the ways to teach math, there's so many different ways you know there's not just one, write it down on the board and they learn it, there's so many different ways that are much more successful as far as teaching students." Mark indicated that he learned mathematics best by watching and copying the teacher and by going back and working through the problems on his own. His belief that he did not learn mathematics best by mastering a series of steps is reflected in his de-emphasis on a step-by-step approach in his own teaching and his efforts to encourage the learners to work through problems on their own. These beliefs were reflected in his instructional practices which were often characterized by an open-ended, exploratory approach to mathematics learning. Toni indicated that she learned mathematics best by "explaining it to others." This appears to be consistent with her ongoing attempts to have learners explain their strategies. While her attempts were not always successful, they provide some indication of how Toni's beliefs about learning shaped her instructional practices. All of the preservice teachers did articulate a belief that mathematics allowed for creativity, exploration, and argument.

One Solution Method vs. Multiple Solution Methods

On questionnaire items that probed the participants' beliefs about teaching mathematics, all of the preservice teachers except Kevin revealed that good mathematics teachers would provide students with a variety of ways to look at a problem. Throughout the semester, the preservice teachers were observed making attempts to provide learners with a variety of ways to look at mathematical problems, including introducing real-world examples and manipulatives. Although they planned for multiple strategies, these were not always evident in their instructional practices. Rather than emphasizing multiple ways to look at a problem, when teaching, they often reverted to showing learners one way to solve a problem. Although Kevin was not certain that good mathematics teachers would provide learners with a variety of ways to look at a problem, classroom observations suggest that he did make attempts to do this when he was teaching. It appears as though his belief may have been shaped, in part, by his interactions with the learners. For example, on several occasions, some of the learners indicated that they found multiple methods confusing and preferred to be given just one right way to solve a problem. Observations of the classes confirmed that the learners were often confused by multiple methods and explanations. Kevin was able to recognize this confusion and satisfy the learners by providing them with a step-by-step procedure or algorithm. All of the preservice teachers indicated that a good mathematics teacher would not show learners exactly how to answer problems on a test. However, they all engaged in a planning session in which they designed a practice test that mirrored the test that was actually administered. During the test review session, the preservice teachers walked the learners through the practice test in a step-by-step fashion. It appears that the preservice teachers were motivated by a desire to have the learners feel prepared and comfortable with the test.

Preservice Teachers' Perceptions of Students on the Margin

Research has determined that secondary mathematics teachers' knowledge, beliefs, and attitudes about learners and student learning have an impact upon the way in which these teachers interact with learners in their mathematics classrooms (Calderhead, 1984). Therefore, this section will examine the preservice teachers' perceptions of students on the margin. With the exception of Mark, the pre-student teaching experience was the preservice teachers' first experience teaching marginalized and disconnected students. Their perceptions of these learners were quite varied. They ranged from being focused on the individual characteristics of the learners themselves to being centered on factors in the learners' environment. For example, while Kevin initially described such learners as "rebellious" and "difficult to work with" and Melissa explained that these learners had "other things that might be prohibiting their learning, whether it be a disability or maybe they have outside family problems." Consistent with his utilitarian view of education, Brad defined these learners as ones who had a high potential to drop out and not finish school or "not get anything out of school that they really need." Because of their backgrounds, Toni and Mark had unique insights into marginalized learners. Toni indicated that, given her own educational experiences, "someone could say I was at-risk." Toni recalled that she "fell through the crack - but it wasn't totally by myself, I was helped, for example, that math teacher helped push me down into that crack." As a result, Toni believed that learners were marginalized through reasons outside their control. Similarly, Mark's two years of teaching mathematics to learners with severe learning disorders and substance abuse problems provided him with unique insights into learners' difficulties. For example, he noted that Jamie's use of drugs and alcohol may have affected her short-term memory and impacted her performance in class. He also stated:

> When you see the student and they're in your classroom, you see this camper and they're in your classroom, all the labels that they've been tagged with don't do justice to their individuality at all so I don't think that, you can use them kind of to say, well this is an at-risk student so people, well, you kind of know there's something going on, but that's about all you can say, there's some, the student has some troubles, uh, some major troubles but you, as far as what those are, and how you can teach that person, no, there's no idea.

All of the preservice teachers voiced a belief that marginalized students had the same capabilities to learn mathematics as learners who had not been labeled "at-risk." Kevin stated, "It's just their lives, their lives have been different prior to this point but I still think they can learn the same, they might be behind now, but I think they can learn just as well." This belief remained constant throughout the study. However, as the semester progressed, the preservice teachers continued to express amazement at the learners' mathematical development and positive attitudes. At mid-semester, Mark noted, "We could be doing a lot more math with these guys." The preservice teachers' enhanced perceptions of the learners encouraged them to change the focus of the course from an arithmetically-focused life skills orientation to algebra and geometry. At the end of the semester, with the exception of Toni, all of the preservice teachers indicated that their perceptions of at-risk learners had changed as a result of the teaching experience. Melissa observed, "They can do the work, it's just that they, haven't been able to do it in a regular school." In one of the final planning sessions Brad offered some additional insights into his perceptions of the learners and the way in which the experience had shaped his beliefs:

> The fact that they have a lot of home problems, and sometimes school's not the number one priority at the time, and so it's tough to reach them, you know, they come there, they're there, but, their mind's elsewhere, you know, depending on whatever happened, and I could see that being a problem at times - they were, a lot better behaved in the class than I thought you know,

> going in, I'm thinking alternative high school and this is going to be terrible and they aren't going to listen to you one bit, they won't do a thing, and then, the reality is they actually did, once you gained trust with them and they gained trust with you, it was, they trusted you to teach them, and they paid attention, and that could be a good lesson, that just any normal teacher going into class, even if you've heard stuff about them, have an open mind, when you first start out to teach them, don't go in with a bad attitude about the student.

The preservice teachers also revealed that the experience had caused them to rethink the label "at-risk." As Melissa explained, "I think a lot of people don't really understand the definition because a lot of people when they think of at-risk they think of students who are bad, you know, students who misbehave." Kevin rejected the label because it did not recognize the many reasons why the learners were at an alternative high school. He noted that some learners were marginalized because personal problems such as substance abuse precluded their full participation in a mainstream high school. However, he also recognized that other learners desired more one-on-one instruction and attention than the mainstream high school could provide:

> I think it, to categorize everybody at the alternative high school as at-risk would be a mistake, I mean, I, yeah, I think it has, some of them have had experiences that haven't been real beneficial but I think, a lot of them aren't so much, at-risk as the ones whose, that's where they've ended up, it's part of their past experiences and, you might find out but, um, there's obviously a couple of kids that were, at-risk, but it wasn't the day to day struggle for a lot of the students, who are at risk for some other reason.

Similarly, at the conclusion of the semester Brad noted, "Even though they're here, it doesn't mean they're not good students and they don't want to learn, it's because they do want to learn, and before I didn't really believe that."

Preservice Teachers' Instructional Practices

Teaching for Procedural Knowledge

Although the preservice teachers' instructional strategies were often student-centered, they focused on procedural knowledge. The practices they employed to facilitate the acquisition of procedural knowledge included emphasizing imitation and encouraging the mastery of procedures and algorithms. For example, lessons that focused on rational number operations involved plugging in values and manipulating symbols. During these lessons, the preservice teachers utilized the expressions, "just set it up", "what's the next step?", and "what would you do first?" A similar approach was used to teach a lesson on area of a rectangle. The emphasis on the procedure appeared to result in the learners focusing on the process rather than the underlying concepts:

Melissa: Ok, each of those boxes on your graph there is 1/4 inch, it's a quarter inch, so if each one is a 1/4 inch, from the sides, then what's the area of 1 of these boxes?
Wendi: ½ an inch?
Melissa: How'd you get that?
Wendi: No clue [laughs] I just guessed.
Melissa: Well, what if I said what is the area of that? [Draws a 3 by 5 rectangle]
Wendi: 15?
Melissa: Ok, so it's 15 inches, or square inches.
Wendi: Square inches.
Melissa: So if this is 1/4 of an inch and this is 1/4 of an inch, you have two 1/4 inch squares.
Wendi: Which is ½.
Melissa: What'd you do to get 15 here?
Wendi: You multiplied.
Melissa: So what do you get when you multiply 1/4 times 1/4?
Mollie: Oh!
Wendi: 1/8, oh, one, 1/16.
Melissa: There you go, 1/16.
Rianne: Oh yeah, 1 times 1 and 4 times 4.
Melissa: 1/4 times 1/4 is 1 times 1 over 4 times 4, so 1/16 square inches, ok? So, now what I want you to do is, looking at this, what can you tell, what do you know the area of, what shapes do you know the area of?
Mollie: That box.
Melissa: This box, so you know the area, you know how to calculate the area of a rectangle, right?

In these instances, the goal of the lessons appeared to be having learners produce a specific response to a problem, respond to closed questions, and find a single, correct answer to well-defined exercises. Although the learners were able to manipulate symbols and provide correct answers to the problems involving operations on fractions, they were not encouraged to provide explanations for the mathematical relations, concepts, and operations. Rather than developing understanding, early lessons which focused on budgeting, stock market, and interest rates, emphasized developing the learners' proficiency with mathematical algorithms. Although the preservice teachers enthusiastically discussed incorporating open-ended, hands-on, exploratory activities into their instruction, in the classroom they often resorted to closed questions, problem solutions, and applications of formulas. The preservice teachers' instructional practices also contrasted with the problem-solving approach to instruction that was the focus of the Methods class in which they were concurrently enrolled. One reason for these differences appears to be related to the preservice teachers' failure to construct lesson plans that specified how the exploratory activities would be incorporated into the mathematics instruction. The majority of the lesson plans written by the preservice teachers contained only brief descriptions of the mathematical topics to be covered and the activities or methods to be utilized. Discussions with the preservice teachers revealed that lack of time prevented them from crafting detailed lesson plans. Additional reasons for differences between planned and implemented instruction will be examined in the next section.

Early in the semester, the preservice teachers seem to have drawn upon their own traditional backgrounds and mathematical experiences when teaching. It appears that the preservice teachers initially designed their instructional practices based upon the procedurally-focused mathematical learning environments in which they themselves had succeeded as learners. The mathematics classroom was not unlike the typical mathematical learning environment, in which a minimal amount of time is devoted to the construction of conceptual knowledge (Borko, et al., 1992; Hiebert and Lefevre, 1986). However, it deviated from the typical mathematical learning environment in that it was predominantly student-centered. As the following excerpt illustrates, learner resistance was one reason for the discussions being procedurally-focused rather than conceptually-focused. When Toni presented a rectangle model to illustrate the concepts underlying the addition of fractions, the learners began mumbling that they did not understand what Toni was trying to demonstrate. As the lesson proceeded and learner resistance to the conceptual focus increased, Toni transitioned to a procedural focus. After drawing a rectangle model on the whiteboard, Toni encouraged the learners to utilize the model to solve a problem:

> Toni: What happens when you add 1/3 of an area to 1/4 of an area?
> Jamie: You get 7/12.
> Rather than accepting Jamie's answer, Toni attempted to have Jamie articulate her strategies and utilize the model:
> Toni: You get 7/12, but how are you getting it?
> Jamie began to recite a procedural approach but then stopped and asked if Toni wanted her to use the model:
> Jamie: Find the common denominator, which is 12, and you take it and [pause] err, are you taking about the boxes or are you talking about this common stuff?
> Toni's response illustrates her interest in having the learners understand the concepts underlying the procedure:
> Toni: Any way you can, all I'm interested in is that you understand exactly what's happening when you divide.
> Jamie: Find the common denominator and then you add [pause]
> Jamie persisted with a procedural explanation which appears to have encouraged Toni to move away from a conceptual focus. While Toni's question may also have been a final attempt to encourage Jamie to explain the concepts underlying the procedure, perhaps by using the area model, Toni accepted Jamie's procedural explanation:
> Toni: Right, and how do you find the common denominator? You're absolutely right.
> Jamie: You find the common denominator by, what number can equal to each other, the lowest number you can equal to each other.
> Toni: Right.

It can also be inferred that, even when they were willing, or intended, to teach for conceptual knowledge, the preservice teachers were not prepared to provide the learners with multiple representations, examples, or explanations of mathematical concepts. Indeed, despite the learners' requests for novel examples of linear functions, the preservice teachers appeared to be unable or unwilling to present the learners with a variety of examples of linear functions. For example, taking a suggested problem from her practicum teacher, Melissa initially brought the pizza problem to the attention of the other preservice teachers: "It costs you $7 to get into a party, and then you're paid $3 for every piece of pizza you eat." Throughout the semester, the preservice teachers repeatedly utilized the pizza problem to teach the concept of linear function:

> Johnny: What are some examples?
> Toni: What are some examples, ok [pause], so, what, what, that's, uh, the basic thing about functions, is that, everybody kind of, could, could you set up a function, if you were given a problem, do you think? (Mollie and Jamie both respond, "Yeah.")
> Toni: Yeah?
> Mollie: Give us a problem.
> Toni: Give you a problem? [Long pause] well, I think we should stick with the pizza one.
> Mollie: Oh, no, no, no pizza.
> Toni: I don't want to make it too hard, because I'm [pause] it won't, the point of what's going on might go [pause] go by the wayside [pause] in favor of the
> Mollie: All right, then give us a pizza one, but not the party one though.
> Toni: Oh, ok.
> Mollie: When you don't know who's coming over, that's, too easy.
> [Long pause]
> Toni: Well, maybe you could think of one, that's, a little easy, so that, it can, become easy to Johnny, maybe that will [pause]

The above excerpt suggests that the preservice teachers' own weak conceptual understandings prevented them from providing the learners with a variety of examples of linear functions. As a result, the learners were unable to identify or define linear functions presented in other contexts. When asked to identify a function, Jamie responded, "It's kind of hard to think of it without referencing it to pizza." During one interview, Toni revealed that she was lacking conceptual knowledge of mathematical content areas that she was expected to teach. She explained, "What I thought I would get from the university math classes was the conceptual background." However, teaching observations and conversations with the preservice teachers suggest that, while most seemed to possess adequate procedural knowledge of mathematical content areas, they appeared to have difficulty utilizing this knowledge to construct conceptually-focused lessons. The preservice teachers also lacked pedagogical content knowledge (Shulman, 1986). That is, they were unable to synthesize their knowledge of teaching with their knowledge of mathematics. As such, they were not always able to provide the learners with the multiple representations, explanations, and examples necessary to facilitate the construction of conceptual knowledge. Perhaps more importantly, they were unable to truly listen to and make effective use of learner comments. During interviews and conversations with the preservice teachers, the researcher noted that they had difficulty articulating a definition, or suggesting an example, of a conceptually-focused lesson. The preservice teachers revealed that they felt pressure from their education classes to implement hands-on, "fun" activities in the classroom but felt unprepared to do so. Toni explained:

> The one thing I've been getting from the Ed [Education] Department is, is, that you know, you come up with those ideas, well, I don't know where to come up with, how to teach the concepts, because I'm not really embedded in the concepts of what is going on, you know, I know, I've never had them, I know the formulas, I know how they work, I know where some of them come from, but I, I haven't, in depth, I've had no experience with tying them to for instance the tearing of corners off the triangle, I love that, didn't, hadn't, didn't know that that was real, you know, I, I never connected that, yeah cool, I like it, and you know that is a great way to, to show them, but they're so afraid of giving out, specific ideas, that you'll take that and go, this is the only way to teach it, that they back off from giving you any ideas.

The preservice teachers did attempt to integrate a variety of conceptually-focused activities into their lessons. These included having the learners (1) explore functions by creating a "starburst" with the graphing calculators; (2) investigate slope on life-size Cartesian axes; (3) utilize gridded transparencies to find area; and, (4) explore volume with "unifix" cubes and boxes of raisins. The preservice teachers also created several activities designed to focus on concepts. These included exploring slope and proportion with birdhouses and utilizing candies in envelopes to teach graphing and linear functions. There were instances in which the preservice teachers, particularly Toni, Mark, and Kevin, were able to successfully utilize these activities to discuss conceptual concepts of mathematics. However, in many instances the potential power of these activities was lost within the lessons in which they were included. The preservice teachers revealed that they utilized procedurally-oriented instructional strategies because they believed that was what the learners needed, desired, and found less confusing. This is consistent with Borasi's (1990) descriptions of the "invisible hand" of students' expectations operating in the mathematics classroom. According to Borasi's research, the expectations of learners' may encourage teachers to employ traditional, teacher-centered instructional strategies. The preservice teachers made attempts to discuss conceptual ideas with the learners. Unsuccessful efforts to engage learners in discussions of concepts appeared to persuade the preservice teachers to revert back to more procedurally-focused instructional strategies. An excerpt of a discussion that took place among the preservice teachers provides some insight into their reluctance, or inability, to teach for conceptual knowledge. The preservice teachers recognized a dichotomy between the strategies suggested in their education classes and what the learners were actually willing or able to do:

> Mark: Hearing stuff in class and then, seeing it, in, when you're actually with students is a totally different level of understanding, just, oh, yeah, it makes sense, when you see it in class, but then you don't apply it to, uh, you know what I mean, apply it to the class is something totally different, that much more, you know, you'd say oh, yeah, I can do that, that's what I should be doing, yeah, and then actually doing it.
> Toni: Actually doing it.
> Mark: This is...
> Toni: Is a different story.
> Mark: Not what...
> Toni: Yeah, well this is supposed to work, it worked in the article. [Laughter]

Moreover, the data reveal that the preservice teachers were aware of one another's unsuccessful attempts to implement group work, investigative activities, or open-ended tasks. It is possible that observations of these failed attempts persuaded the preservice teachers to utilize more closed-ended, traditional instructional methods in order to prevent a perceived loss of control, or even chaos, in the classroom. For example, early in the semester, Toni attempted to teach the addition of fractions with unlike denominators for understanding by utilizing a rectangle model. Her own weak understandings were revealed during her attempts to utilize this strategy in the classroom. As a result, both Toni and the learners becoming confused. Kevin then provided the learners with the formula which appeared to satisfy their need for an algorithm and alleviate some of their confusion. As the teaching centered more on the procedural aspects of the operations involved, the focus of the lesson changed from conceptual to procedural. These tendencies will be examined in more detail in an upcoming section. The following is one example of the preservice teachers' attempts to design a student-centered, conceptually-focused lesson that included collaborative group work, open-ended discussion, teacher facilitation rather than transmission, and multiple problem solving strategies. The discussion began with the preservice teachers planning to put the learners in groups and have them collaborate to explore and discover ways to find the area of a triangle:

> Mark: Um, [pause] well I know that they should know basically how to, find the area of a triangle, and that's all you really need, the bare minimum to do these, it won't be hard, um, so I'd say just throw them out there in groups and let them work on that.
> Toni: But are you going to tell them, use the triangles, the area of a triangle, or let them?
> Kevin: Let them.
> Toni: Let them figure out that they can do that?
> Kevin: Let them be as creative as they want to be.

Mark: One group starts estimating.
Toni: Yeah.
Mark: That's one method, let them do it.
Toni: So, so, the, the, amount of instruction, that, that, you give them, is just take these shapes, and [pause] find the area?
Brad: Right.

Kevin and Mark agreed that, instead of providing the learners with formulas, the preservice teachers would offer the learners guidance or assistance. In addition to encouraging learner explanations, the preservice teachers recognized the possibility that multiple methods would emerge from the learners' explorations:

Kevin: Right, and I think that's with minimal help.
Mark: Yeah, and if, I'm sure someone will, will be on, a good track, of a good method, and I think probably all of them will, if none of them happen to find a good method, then, uh, after they've worked out their answers the way they think they should be worked out then we can go over different methods, theirs and others.
Brad: I say we have them explain, like if one of them did something different than the other, have them explain.
Mark: Have them explain it to each other.
Kevin: And ask them, maybe, oh sorry.
Brad: It's ok, I was done.
Kevin: If uh, they do come up with different methods, an interesting one would be, say, ok, "which one's right?"
Brad: Or they're both right.
Mark: Get into some discussion.

As the planning continued, the preservice teachers indicated that they desired to deemphasize formulas and single solutions. Instead, they planned to discuss and recognize multiple strategies and methods:

Kevin: Whether there's a right way or wrong way, cause that's, I find a lot of them, are really stuck in what's the right way to do it, what's the wrong way, how do you do it, what's the formula.
Toni: So, so…
Kevin: And just plug it in.
Toni: So maybe when you tell them take these shapes and find the area, you should, you should also tell them be able to explain your answer, cause I know a lot of them start to clam up, they're, two of the, start to clam up the minute they, have to, explain their answers, and they uh, ahead of time, saying you have to be able to explain your, how you got that, how you found this.
Kevin: Yeah.
Mark: Well, I'll just say, I want to know, how you got your answers.
Kevin: They should have to present.
Toni: Yeah, yeah.
Kevin: In the front of the room, how they got their answer.

At a subsequent planning session the preservice teachers sought to have the learners discover, explain, and discuss their methods for finding the areas of different shapes:

> Kevin: Say, everybody pick the most complicated one to find the area for...and then have them show you how, like come up to the front of the class, show everybody how they got the area for that, and then, start talking about that, well, I don't know, I think that could get them really involved.
> Toni: Yeah, yeah.
> Kevin: As opposed to lecture.
> Toni: I, I definitely want to tie them into their, to their, um, floor plans, and the shapes, that they have.
> Mark: So have them figure out their own methods for calculating areas and then, re-in, pull it together, I think that would be cool.
> Kevin: Yeah.
> Toni: Mmhmm
> Mark: Then you, I think, you'd probably want to go, you might want to go general, how they all did, and then maybe go with one of them, before you go with your own.
> Kevin: Maybe just have everybody present, then talk about it, so that way you could...
> Mark: And when you talk about it, would you want to draw it towards a general theme, or would you want to draw it towards a circle so that later that day you could do the circumference? A more precise calculation.

Both Mark and Melissa later indicated that they had attempted to encourage learner discussion and exploration of the underlying mathematical concepts, including area, measurement, and geometric properties. Mark began the lesson by asking the learners to articulate strategies for finding the areas of polygons when they didn't know the specific formulas. During the course of the discussion, Mollie and Johnny recognized that they could break a polygon down into shapes whose areas they knew how to calculate:

> Mark: So, um, what are some things that, how have we done this before, when we didn't know the formula that we could just plug in? What are some different things you've been doing? [Pause] you don't remember?
> Melissa: What about when you have this shape?
> Johnny: I used a bunch of triangles.
> Mollie: Oh yeah, we made it into different shapes.
> Melissa: What shapes did you use?
> Mollie: Triangle and rectangle.
> Melissa: Right, so, can you think of a shape that you might be able to calculate the area with?
> Mollie: Oh yeah.
> Mark recognized an alternative strategy utilized by Johnny:
> Mark: Johnny has another idea of using the uh
> Johnny: I don't know what it's called.
> Mark: Using the graph paper transparency.
> Mollie: Yeah.

Mark continued his attempts to encourage the learners to articulate their strategies:

Mark: You can do it that way [pause] how did you get it? [To Rianne]

Rianne: Uh, first I measured these, from one of the sides.

Mark: Mmhmm.

Rianne: [showing Mark] and times-ed it, and it came out 6, and I did, I measured the, different sides, you know, just, like that, times-ed, divided it by 6, and then you could do it another way.

Mark: So you took the, area across, and then you took the area this way, from the line.

Rianne: Mmhmm.

Mark: And squared it, and then you times-ed them, do you see what that's going to give you?

Rianne: What?

Mark: Like if you drew these like this, and then you multiplied them, that would be the area of that square, length times the width, so that will give you, that area, and you found that area, by multiplying that by that, and then you divided by 2? [Pause] one thing you could try is, uh, doing it some way that we talked about before, when we talked about…

Rianne: [inaudible] Divide by 2?

Mark: That's the area of a triangle.

Rianne: Oh, that's just the triangle.

Despite Mark's efforts to engage the learners in a discussion of the underlying concepts, they appeared to remain focused on the procedures. While the learners were able to articulate their procedures, they struggled in their attempts to explain their strategies and their mathematical understandings during the lesson. However, they did exhibit spatial sense and an understanding of area in terms of its deconstructed parts. Moreover, Johnny demonstrated understanding of area in terms of a unit. At the planning session which followed the above class, Mark recognized the learners' struggle and expressed frustration at the unwillingness or inability of the learners to explore the concepts. In the following transcript, the lesson to which Brad refers is discussed in more detail later in this chapter:

Mark: So last time when I tried, I tried just having, giving them a problem, having them work it out, and they just don't, they just don't, they weren't clear and refused to think it out, and

Melissa: Which one? Oh, you mean the

Mark: They just don't, if it just doesn't click, Mollie does, Mollie will start thinking of ways, but everyone else is just, I can't do it, and they'll just sit there.

Melissa: Which one are you talking about?

Mark: When I was having them figure out the area of the polygons

Brad: Yeah, that was the same way when I was helping them do like the "n minus 2 times 180," or whatever, she [Mollie] was the one that actually sat down and worked, the other people just said, "oh, I don't know."

The above data suggest that the preservice teachers experienced frustration with the learners' unwillingness or inability to engage in discussions of the mathematical concepts. Similarly, the learners made several comments which suggest that, while they were aware of the preservice teachers' efforts to teach for understanding, they preferred formulas which were "easier." The following exchange took place during a lesson in which Toni tried to teach the concept of area for understanding:

Johnny: Now wouldn't it just be easier if you multiplied?
Toni: Yes it would be easier if you multiplied, it would
Johnny: You just wanted to show us how it works, right?
Toni: Right, right, yeah.

The preservice teachers were aware of the learners' discomfort with and resistance to open-ended activities. Their desire to maintain a positive learning environment and bolster learners' self-esteem may have motivated the preservice teachers to utilize more traditional methods of mathematics instruction. Other data, specifically interviews and discussions with the preservice teachers, imply that they were uncertain about whether, and when, to teach conceptually. For example, after observing a video clip of the learners struggling to respond to one of his lessons, Mark observed, "The retention isn't there, I think they needed more practice, and less concepts." However, later in the study, Mark stated, "I mean, if I were to do it, then, I don't know if it would work better or not, but I'd try and do more concepts and then the formulas come in at the end, not at the beginning." The preservice teachers' discussions also revealed their differences about whether and when to emphasize concepts or procedures. In order to provide the learners with an innovative way to explore the concepts of area, volume, surface area, scale, and cost function, the preservice teachers had each learner create a scale model of their "dream dwelling." However, the preservice teachers were not in agreement about how to utilize the models to explore these mathematical concepts. For example, Kevin initially believed that the mathematical concepts would emerge as the learners constructed their models. In contrast, Toni was doubtful that the learners would be able to discover or visualize the concepts and suggested that the preservice teachers explicitly discuss the concepts as the learners worked on the models. In a subsequent planning session, after the learners had spent several class sessions on the models, Kevin said, "I kind of thought it would be good, this is my opinion, to, to be teaching the concepts as we go, and then the applications later." Other examples of unsuccessful attempts at engaging learners in conceptual interactions emerged. The following transcript is of a lesson Mark taught on circles:

Mark: You guys remember how to find the area of a circle?
Mollie: Diameter times pi.
Mark: You guys remember that equation? [Pause] so for the circle, um, what's the diameter? What'd you get?
Mollie: 2 inches, it was a little less than 2 inches but
Mark: Ok, you guys remember what pi is?
Mollie: 3.4, 3.14.
Mark: 3.14, and so...

Jamie: So what's 3.14?
Mark: That's pi, that's just a number, and when you have a circle
Jamie: Where are you getting that number?
Mollie: It's just some number, a number somebody made up.
Mark: Um, were you here that day when we were measuring, the length around the outside? If you take a, like, um...
Mollie: I went and measured the clock, I measured my peanut butter cup.
Mark: This is, [pause] so if you take this, tape measure, around the outside like that, remember that? And you measure that length, whatever that is, that'll give you a number, and it's, um, this distance here, uh, which is 2 times some number, say x, is going to be equal to that length there, um, and this number is pi, it's just what works out.

As the discussion continued, it assumed a procedural focus. Although Mark attempted to have the learners discuss the concepts underlying the formulas, the learners were more focused on talking about the formulas. Jamie then posed a question which shifted the discussion back to a more conceptual orientation:

Jamie: Ok, but you're not doing that, you're, you're trying to figure, I understand that that's what you're trying to do, but I don't understand how you're getting that number.
Mollie: 3.14? That's pi, that's pi.
Mark: Yeah, that's the same as pi.
Mollie: But she wants to know, where is pi coming from?
Jamie: Yes.
Mollie: Why is pi, 3.14?
Mark: For any circle, when you measure around, the distance around, is, 3.14 about, approximately 3.14 times the distance across.
Jamie: But if you have a circle this big, how can it be 3.14? [Makes a huge circle with her arms]
Mollie: Because, look at this, we did this, remember? You did it too, you were here that day, and we measured the caps and they all came out to about 3 point something.
Jamie: All right, all right, all right, yeah, yeah.
Mark: As the circle gets bigger, the circumference and the diameter both get bigger.
Jamie: Oh.
Mark: And then for area, this distance here, times itself, square it, and times that number by pi, 3.14, that's how you get the area [long pause] um [pause] you seem a little like you don't like that, or, do you have a question?
Jamie and Mollie both respond, "No."

Despite having "discovered" pi and working with the related formulas during the previous class with Melissa, the learners were struggling to fully participate in and respond to Mark's lesson. It appears that the learners lacked knowledge of both the underlying concepts and the procedures or formulas. Mark's comment again raises the issue of how the rotating teaching schedule affected the continuity of the teaching. As he read the transcript of this lesson, Mark observed that the learners had minimal knowledge of the concepts and the procedures covered during the previous lesson. As a result, Mark was uncertain about whether to focus on concepts or procedures:

> Basically, the problem I'm having is the students don't really know this, the formulas, that they, the concepts that they've already covered, so it's, do I cover the concept or do I just show them the formula, which they don't seem to know at all, they don't know anything about, so, that's part of the problem when you go, switching days, you don't know what's been covered, really, and I'm not doing any concept teaching in this, this is, this comes in the middle of the lesson, I don't know, I probably did something before, but this is just, this whole thing is trying to understand the area of the circle, um, where the area of the circle formula comes from, which they already did earlier, [pause] it's hard to say, I didn't want to go into that, but I couldn't avoid it.

The preservice teachers could not distinguish, and were not always in agreement about, the differences between applying formulas and solving problems. This may account for their confusion about what distinguished a conceptually-focused lesson from a procedurally-focused one. An example of this appears in the following excerpt, which is taken from a planning session in which the preservice teachers were planning a written test for the learners:

> Toni: Ok, now what I want to get into this exam is some, is some, manual, equations, and stuff, right? You know just, solve the equation.
> Mark: That's what this is right here.
> Toni: Which one?
> Mark: This area stuff.
> Toni: This? Well that's not actually solving, that's, well, I guess, I don't...
> Mark: It's just taking the different shapes and applying, the formulas.
> Toni: Yeah, this is applying formulas, it's not actually solving, is it? I guess it would be solving an equation, I was thinking about just giving, you know, solve this equation, you know, hypothetically.
> Melissa: This is solving for an equation though.
> Toni: Yeah, that is, you know, given, given, you know, and x, and you know [pause] that equals whatever.

Throughout the semester, the preservice teachers continued to struggle with developing conceptually-based teaching strategies. While Toni recognized that she possessed only a superficial knowledge of many of the mathematics concepts she revealed that, as a full-time university student, she did not have the time to explore the concepts in depth the way she would when she was teaching full-time:

> When I become a full-time teacher and a no time student, I will definitely spend way more time, uhm, researching the concept that I'm teaching that week or that day, whatever, it, it it's my nature to do that anyway and it just, I just really have a hard time with the fact that I don't have the time to do that, unless they change the whole concept, then I'd just have to review on it and

> whatnot, and I've had to, with my math education, had to just get little tastes of these things and have had no time to really go in depth into where each concept can go, so I'm I'm basically on several subjects on the surface of it, you know, and it bothers me that I'm just on the surface and I want to go further, but you know, as a student, you've got to go with what you've got.

Although the preservice teachers were able to teach in conceptual realms on occasion, the majority of the instruction was procedurally-focused, but student-centered. There were several reasons identified for these tendencies: (1) learner resistance and lack of knowledge; (2) lack of direct guidance on how to implement "conceptual activities"; (3) preservice teachers' confusion about what a conceptually-focused lesson is; (4) preservice teachers' beliefs and attitudes rooted in their own educational experiences; (5) their initial perceptions of marginalized learners; and, (6) lack of time to prepare lessons and reflect on the concepts underlying the mathematical topics in those lessons.

Teaching Conceptually

Many of the previous examples of procedurally-focused teaching cited in the previous section contained some aspects of teaching for understanding. However, while these lessons were student-centered, they did not delve into conceptual realms. In addition, many of the lessons described in the previous section were either planned or began as conceptually-focused lessons. However, for reasons previously given, these lessons took on a more procedural focus. Although much of the instruction was procedurally-focused, there were instances in which the preservice teachers successfully engaged in teaching for understanding and conceptually-focused instruction. This section will explore lessons which had a focus on the conceptual underpinnings of the mathematics being discussed but were not student-centered, as well as lessons that were both student-centered and conceptually-focused. Additional examples of conceptually-focused instruction are included in the case studies of the individual preservice teachers. In particular, Kevin's lessons on scale and angle measure, Toni's lessons on area and ratio, scale and proportion, Brad's lesson on slope, and Mark's lesson on linear functions.

Technology Tools. All of the preservice teachers indicated that they were very comfortable and experienced with various technology tools. In addition to utilizing standard, hand-held calculators on a regular basis, the preservice teachers also used the Geometer's Sketchpad, graphing calculators, and *Excel* spreadsheets. However, the preservice teachers' comments suggest that they viewed technology as a skill to be mastered rather than as a tool for exploring mathematical concepts. Mark suggested, "Teaching them the technology that they can use to help them out." Toni and Kevin extended upon this view of technology:

> Toni: Because you could extend the life skills into business skills, bring in some technology, have them learn the graphing calculator or some computer time

to get familiar with a math program on the computer - just to get them used to using computers too, and they could use those to do their banking, whatever...make their budget record - I'd definitely like to see the technology worked into it some kind of way because that's today's real world - if you don't know anything about technology, you can forget a high paying job - you've just got to be aware of it if not proficient at it.

Kevin: And that would give them a skill they could take too.

Toni: Yeah, a skill they could benefit from when they go to find a job - it would put them at the upper end of the lower end jobs where there's so much competition anyway, technology skills, a priority.

Out of a total of twenty-six classes, the preservice teachers made significant use of technology in only four lessons. Classroom observations of these four lessons revealed that Toni was the only preservice teacher who was successful at utilizing technology to explore mathematical concepts within a student-centered learning environment. During two lessons, Toni introduced the graphing calculator and attempted to have the learners utilize these to enhance their understandings of the concepts of function and slope. For one activity, Toni had the learners use the graphing calculator to create a "starburst" design. All of the learners appeared to be engaged and were able to follow Toni's directions and manipulate the buttons with ease. For the most part, they were successfully able to reproduce the "starburst" on the graphing calculators. Toni made efforts to engage the learners in a discussion of the underlying concepts. The learners were engaged and could have easily been reflecting on slope, intercept, or shift:

Toni: Let's see what you got, push graph, all right, you've got 2 lines, ok, so, what's the difference between those, well, what's going on there, what's the difference, what made one line change? You changed one thing, what, what was it? Ask yourself, what's the same and what's different? [The learners did not respond but continued graphing]

[The learners continued graphing. In an effort to encourage the learner to explore fractional slopes, Toni posed another question] How come you, how come you, what do you think is the reason those came in that area?

However, as Toni drew the lesson to a close and attempted to solicit discussion of the learners' strategies for creating the "starburst" design, the discussion evolved into a lively conversation about the origins of mathematics. Despite the learners' reluctance or inability to participate in a discussion with Toni, a good deal of learning took place. For example, the learners, especially Mollie, demonstrated knowledge of the properties of slope and y-intercept as well as connections between equations and graphs. The instructional practices utilized during the only lesson that incorporated the Geometer's Sketchpad mirrored the general pattern of instruction. For example, the preservice teachers walked the learners step-by-step through a series of activities. Kevin, however, attempted to have the learners explore an underlying concept by engaging them in an activity that required using Geometer's Sketchpad to construct triangles between two parallel lines. As they manipulated the triangles to change their areas, the learners were encouraged to observe that the interior angle measure remained at 180 degrees. Although the preservice teachers provided the learners with a great deal of one-on-one assistance, there was no whole-class discussion of the concepts explored with the Sketchpad. At the planning session which followed the Sketchpad, Kevin recognized that he had guided the learners through a series of steps:

> Well, it was, was more me showing them, say divide this by this, and so, I mean they, I told them what to do because I was in a hurry, because we were out of time, and I thought it would be a neat thing to do on the Sketchpad [pause] so they saw it, but they didn't discover it, no, if you know what I mean.

The lesson on the Microsoft *Excel* provided the learners with a brief introduction to this spreadsheet program. After receiving a cursory explanation of the software, the learners spent the rest of the lesson entering their budget data into the spreadsheet. Once again, the preservice teachers offered individualized, one-on-one assistance but did not engage the learners in a whole-class discussion of the software and its applications. This practice may have adversely impacted whole class interactions because the explanations and strategies were not shared with all learners. The Sketchpad and spreadsheet were not utilized again nor were the learners encouraged to use the technologies to explore mathematical concepts. This was unfortunate because observations and discussions with the learners revealed that they were quite comfortable and enjoyed working with computers. Discussions with the preservice teachers revealed that their education classes did not provide them with a knowledge of more creative ways to introduce and integrate a variety of technologies into mathematics instruction. The preservice teachers were also reluctant to allow learners to explore with the technologies because they were worried about possible classroom management problems. While the excerpt that follows concerns graphing calculators, the preservice teachers expressed similar sentiments about the spreadsheets and the Geometer's Sketchpad:

> Brad: The only thing about graphing calculators, teaching it, I see this as a problem in a real classroom too is we have to make sure they aren't just screwing around with them.
>
> Toni: Yeah, well, I thought the way that I would approach it is, is introduce the buttons that we're going to be using and then I'd have them use some of the equations they've been working with over the past couple of week and have them graph that.
>
> Brad: We're just going to have to, everybody that's there, we're going to have to make sure they're listening to the teacher and not just screwing around with the calculators.

Collaborative Learning Groups. Collaborative learning groups were tried on one occasion and then abandoned as an instructional strategy. According to Toni, an "unspoken friction" between learners discouraged group work. The preservice teachers' comments reveal how their instructional practices were impacted by their perceptions of the learners as well as by learner resistance:

> Melissa: Group work didn't really work too well cause they kind of pretty much wanted to do it independently.
> Mark: They were totally opposed to the idea of group work.
> Melissa: We tried though.
> Toni: Yeah, we tried.
> Melissa: We tried all the things we learned, you know.
> Mark: Students do like to do group work, it's just our students didn't.]
> Toni: And that probably was tied into their personal problems that they had experienced, either, uh, I know one particular student had great trouble with the chaotic, uh, um, nature, of, of group, when, when, when we would all started actively bouncing about the room and participating, uh, you could just see the, oh, my god, I can't deal with this come over her, and she would go sit down and go, huh, huh, you know, she didn't, visibly do that, but, you could tell she was like, and, and, so she just was not comfortable with that, and, uh, yeah.
> Brad: When they were in the normal school there had been group work and they had been left behind, you know, like if you don't understand, the group just goes ahead and does everything, and so I think that kind of.
> Toni: Shut 'em down.
> Brad: For at least one or two of them, if not all them, it kind of turned them away from group work.

Use of Student Comments to Facilitate Discussion. Discussion is defined as purposeful talk on a mathematical subject in which there are genuine learner contributions and interactions (Pirie, 1988). In order for this sort of purposeful talk to occur, learners must be allowed to talk with their peers, argue, experiment, invent and justify alternatives, and be wrong (NCTM, 1989). The preservice teachers attempted to engage the learners in discussion throughout the semester and each stated that as an instructional goal. When the preservice teachers reflected on their lessons by observing video clips of their teaching, all indicated that they had attempted to facilitate discussion of the mathematical concepts. Toni's comments are representative of what the group of preservice teachers aspired to do as they planned for the semester:

> I would definitely encourage student explanations, having students that felt comfortable with the material, rephrasing it for, for students who weren't comfortable with the material, and, and then have, I think I would try to have the students that were uncomfortable, unsure of the material, try to explain it to the ones that were comfortable to see if they could get their understanding across to students that, felt they had a good understanding, because even if you feel you have a good understanding of a concept, you still might be missing some subtle point that someone with a, simpler outlook on it might connect with, and, and then you go oh yeah, I wasn't even thinking that, so, I think that's how I think I would do discussion and then, you know, ask some prompting questions of the students involved, or something, it's hard to be specific about actually [pause] doing it.

The following transcript excerpt provides an example of a lesson in which the learners were engaged and the preservice teachers were focused on understandings of the multiplication of negative numbers. This particular segment also illustrates the potential power of team teaching as the preservice teachers involved in this lesson were able to provide the learners with multiple explanations for the mathematics being taught. In this lesson, Mark and Toni attempted to teach the multiplication of positive and negative numbers with understanding by giving the learners word problems and using money as a context for a conceptual model. However, their lack of pedagogical content knowledge prevented them from providing the learners with a word problem that represented the multiplication of two negative numbers:

> Mark: Yeah, minus 2 times minus 7 you have 2 of the opposite of minus 7.
> Mollie: How can that be? If you have negative 2 dollars and you owe 7, how can you end up with 14?
> Mark: Oh, that's just a little different, uh, let's start up here.
> Mollie: How would you put that, how would you put that into it?
> Toni: Well, the 2, you would, if you owed 2 negative 7's, the 2 would be positive, you'd be in the hole 14, negative 14.
> Mark: Let's try it this way, if you have two $5 bills, right? That's $10, if you have uh, if you owed two $5 bills, that becomes minus 5, then you owe $10, right?
> Mollie: Mmhmm.
> Wendi: Mmhmm.
> Mark: Now if you...
> Toni: Didn't owe...
> Mark: Hold on, [long pause] [writing]
> Jamie: That negative and negative, how do you get a positive?
> Mark: This, this is the same right, basically, this says you have [pause] this is like you owe two people $5, this first one, you owe two people $5...

Because of Toni and Mark's inability to utilize a conceptual model for the multiplication of two negative numbers, the learners were again confused. As in the previous lesson, Kevin was able to provide an explanation which seemed to satisfy the learners, "I would say if you did negative 2 times negative 5, I would say, you didn't spend $5 twice so you have $10." Kevin drew upon his knowledge of Wendi's background and interests to provide her with an alternative explanation: "Wendi, you're interested in like English and if you had written a sentence and you put negative words in there, then that would equal positive, f you say I'm not not going to be there. That's gonna mean you are going to be there, it's the same thing."

Despite their desire to engage the learners in mathematical discussions, it appears that the preservice teachers were, for the most part, not prepared to consistently utilize student comments to facilitate discussion of mathematical concepts. While all of the preservice teachers indicated an intent to facilitate discussion, they revealed that, because they were simultaneously focused on a variety of other teaching issues, they were not always able to recognize and react to student comments. Their lack of experience orchestrating whole-class discussions of mathematical topics may also have inhibited their ability to engage learners in discussions. However, there were many instances in which the preservice teachers successfully orchestrated conceptually-focused discussions. The following transcript excerpt provides an example of a student-centered, conceptually-focused lesson in which learners were engaged in a discussion of mathematical concepts. Additional examples of conceptually-focused lessons are presented in the case studies of Kevin and Toni. In the following transcript excerpt, Mark began a lesson by laying a variety of paper polygons out on the table. He then asked the learners if they could group any of them together. During this lesson, Mark sat at a table with the learners. It should be noted that Rianne, a student who is normally extremely taciturn, responded to Mark's comments. Throughout the lesson, Mark encouraged the learners to articulate their observations.

Rianne: Yeah, that one is all funky, and that one's perfectly, every side is the same.
Mark: Every side's the same on this one, any one, any of the others have all the sides the same?
Rianne: This one.
Mark: This one, yeah.
Rianne: That one.
Mark utilizes Rianne's comment to extend the discussion:
Mark: And that one, yeah, those three, these three are called, uh, they're called regular shapes, or regular polygons, a polygon is just, more than, 3 or more sides, and regular means that all of the sides are the same, um, now this one's funky looking, right, is that what you said?
Rianne: Yeah.
Mark encourages Rianne to think about the underlying concepts by posing open-ended questions:
Mark: Um, why? What is it about it that makes it look different?
Rianne: It's got a lot of sharp points.
Mark acknowledges Rianne's comment and encourages the other learners' to articulate their observations. This prompts Jamie to extend the realm of geometric properties under discussion:
Mark: Sharp points, yeah, anything else that you can notice?
Jamie: A lot of acute angles.
Rianne: On that one.
Mark: Yeah, acute angles.
Jamie: Is it still a polygon?
Rianne: Mm, yeah.
Jamie: All the sides are different, some of them.
Mark utilizes Jamie's question to extend the discussion:
Mark: Good question, some of you have, some of the angles go in, the other way, right? Instead of out, all these ones stick out, point out, some of these point in, or some, some of you are looking at the inside, it's bigger than 90 in here, bigger than 180 in there, because this one's bigger than 90.

Rianne: Uh huh.
Mark: And this one's bigger than 180.
Rianne: We were talking about that.
Mark has attempted to clarify the properties of various polygons by attending to angle measure. He then attempts to encourage the learners to connect the concepts in the current lesson to those discussed in a previous lesson:
Mark: Huh? You already talked about that? So you guys remember figuring out some ways to figure out the triangle has 180, remember doing these at all, seeing different shapes?
Rianne: Mmhmm.
Mark: Like this one.
Rianne: We made our own shapes.
Mark encourages the learners to describe their strategies:
Mark: And how did you figure that, how did you figure it out?
Rianne: We measured each angle.
Mark: You measured each angle.
Rianne: Of the shape, that we created, and added them.
Jamie: We have an easier way of doing it.
Rianne: Yeah.
Mark recognizes the potential of engaging Jamie in the discussion so he continues his strategy of utilizing open-ended questions to encourage the learners to discuss concepts:
Mark: How's that?
Jamie: Every side, every, I mean they have more than the, if every side, in the polygon, it's more than 180.
Rianne: Yeah.

Mark employed several strategies in his conceptually-focused lesson. Throughout the lesson, he utilized student comments and responses to extend the discussion of the mathematics concepts. Mark also posed open-ended questions and encouraged the learners to articulate their observations and strategies. He also made several attempts to encourage the learners to connect the mathematical concepts in the current lesson to concepts discussed during previous lessons. According to the National Council of Teachers of Mathematics' *Professional Standards for Teaching Mathematics*, in order to encourage learners' reasoning about mathematics, a mathematics teacher should consistently follow learners' statements with, "why?" or by asking them to explain (NCTM, 1991, p. 35). However, the preservice teachers rarely utilized "why" questions or asked the learners to explain their answers or justify their comments. Moreover, the preservice teachers' often transmitted both conceptual and procedural information directly to the learners. Perhaps because of learner resistance, the preservice teachers appeared reluctant to pose questions that challenged the learners to justify their statements. Classroom observations revealed that the preservice teachers were more likely to teach geometry conceptually and were more likely to teach arithmetic and algebra procedurally. This difference may be due, in part, to the fact that the preservice teachers were able to provide the learners with multiple examples, discuss underlying geometric concepts, and explore connections between concepts. As the semester progressed, the preservice teachers' perceptions of the learners' mathematical abilities and learning patterns were enhanced through their interactions with the learners. It appears that these positive perceptions encouraged the preservice teachers to attempt more conceptually-focused activities which challenged the learners to reflect on and discuss the underlying mathematical concepts. As the semester progressed, the preservice teachers became more comfortable utilizing investigative and student-centered activities. It appears that, as they acquired experience teaching mathematics, the preservice teachers appeared more at ease asking open-ended questions and orchestrating conceptually-focused discussions. The data suggest four reasons why the preservice teachers were, on occasion, able to successfully teach conceptually. These reasons were identified as: (1) mathematical topic; (2) source of activity; (3) enhanced perceptions of learners; and, (4) acquisition of mathematics teaching experience.

Case Studies

In the three sections that immediately follow, case studies will be presented for Brad, Kevin, and Toni. These preservice teachers were selected for case studies, not only because of their diverse backgrounds and beliefs, but also because of their distinct instructional practices. While the majority of Toni's lessons were characterized as teaching for understanding, Brad's lessons were primarily procedurally-focused. The bulk of Kevin's lessons were dominated by two different instructional practices. His lessons were either conceptually-focused, without significant learner involvement, or procedurally-focused with a great deal of student involvement. The case studies will focus on the relationships between the preservice teachers' individual beliefs, perceptions of at-risk learners, self-concepts, and instructional practices. In the last two sections, abbreviated profiles will be presented for Melissa and Mark. These profiles will focus on their individual beliefs, perceptions of at-risk learners, self-concepts, and instructional practices.

Brad

Brad was the most reserved and reticent of the five preservice teachers. He was serious, disciplined, and did not engage in verbal bantering or jokes with the learners. During the course of the study he would often catch the researcher's eye and shake his head or roll his eyes when he disapproved of a behavior or comment expressed by a learner or another preservice teacher. As a result of his positive experiences with both high school and college mathematics courses, Brad entered the study with a high mathematics self-concept. Because he did not have any prior teaching experience, and did not speak about his confidence level at the onset of the teaching experience, it was not possible to assess Brad's teaching self-concept at the beginning of the study. However, interview and observation data suggest that Brad was initially concerned about issues of classroom management. Brad indicated that he was somewhat anxious and worried about what to expect from the learners. Brad revealed that, "I think we were all kind of worried about the behavior of the students and how they'd act." Brad explained that he continued to take mathematics classes because he was both interested in the subject matter and "just good enough in it to keep going." Brad's teaching was characterized by a serious, no-nonsense style. Even Wendi, the most unruly learner, recognized early on that Brad had little tolerance for teasing and rowdy behavior in the classroom. Although the other preservice teachers' had varying degrees of difficulty with Wendi she rarely acted up when Brad was teaching. Brad expressed a belief that success in mathematics was a result of a propensity to do math. He defined talent, not as a natural aptitude to do math, but as a willingness or desire to do math. As he phrased it:

> I think it's talent of the fact that some people like doing math, as opposed, I mean, and also well it goes along with when you like doing math you get so much better at it and you continue liking it, a lot of people I know haven't liked math forever and so they don't, they avoid it as much as possible, so I'd say it's more of a talent of [having a] willingness to do math.

Brad's view that success in mathematics was a result of a willingness, propensity, or desire to do mathematics remained constant throughout the study. However, at the conclusion of the teaching experience, Brad broadened his belief to identify success in mathematics as a result of "hard work, because I believe that anybody can be successful at it." Brad's belief that all learners had the potential to do mathematics also remained constant throughout the study. He indicated that, because success in mathematics was a matter of a learner having confidence, learners who underachieve in mathematics are "the ones that aren't confident within themselves and don't believe in their ability." Brad also stated that a lack of interest or desire to do mathematics prohibited learners, noting, "If they get confused or troubled at the beginning they basically shut off math." At the final interview session, Brad illustrated his beliefs in the context of a specific mathematical content area by describing what he thought was easy and/or difficult about the concept of slope for learners. He once again cited confusion as having a negative impact on mathematical development:

> I think past experience with the concept helps, and also how the teacher teaches it, if the student gets confused right away then they might kind of shut out to learning it whereas if all of a sudden if some teacher says something that clicks in them right away then they'll understand it better.

Brad's teaching style appeared to be consistent with his expressed belief that the most important reason for teaching mathematics is to provide learners with mathematical skills. As such, Brad's lessons were orderly and methodical. Much of Brad's teaching was procedurally focused and teacher-centered. Although the learners accepted Brad's style and responded to his lessons, there was scant evidence of enthusiasm or spontaneous discussion.

Brad: How about changing fractions into decimals, anyone have problems with that? [Long silence]
Mollie: Top divided by bottom [pause] right?
Brad: Mmhmm. OK, walk me through, what's the first step? Anybody?
Jamie: Put the 8 in the little box and the 5 on the outside.
Rianne: Oh, yeah.
[Brad set up and worked through the problem step-by-step on the board]
Brad: And then what?
Jamie: 5 goes into it once, 3 leftover, 5 goes into 30, 6 times.
Brad: And what's this 1.6 as a percentage?
Jamie: 160%
Brad: No one has problems with that, taking decimals, or fractions turned into decimals, decimals to percents, or the other way?
Jamie: You could have done that easier.
Brad: What?
Jamie: Just, 5 into, 5 goes into 8 once with 6 leftover.
Mollie: I know an easier way... calculator. [Laughter]

Brad's instructional practices may have been influenced by his utilitarian view of mathematics, "Just because math is prevalent so much in everyday life, even though people do not realize it, that it is, and, without math skills, you're, there's no chance really for a person to succeed." It appears that Brad's beliefs and instructional practices were influenced by his prior experiences as a student in traditional high school and college mathematics classes. He excelled in mathematics throughout high school and college and indicated that he felt comfortable with both the teachers and the instructional strategies. Brad identified his knowledge of mathematics as one of the strengths of his teaching. However, he identified as a weakness his difficulty in articulating his ideas in a way that would make them understandable to learners. While this contrasts with student-centered, investigative teaching strategies, it is consistent with Brad's procedurally-focused approach to instruction:

> The weakness is that sometimes I don't know how to say it in words where they will understand it, because I think of, a lot of math problems, I think of as different than they're explained in books or in classes, and it's hard to explain my ideas to other people, if they're trying to look at them say the way they've been explained to them before, or the way the book says it, or my strength I would say is that I know what I'm talking about, but the weakness is I have trouble verbalizing it.

It appears that Brad developed his own teaching style by modeling the traditional teaching styles he observed in his high school and college mathematics courses; courses in which he was extremely successful. A review of videotapes and transcripts of Brad's teaching suggests that he utilized direct instruction. Direct instruction assumes that the teacher acts as a transmitter, imparting knowledge to learners through modeling and step-by-step instruction (Goldenberg, 1992). The learners' willingness to participate and respond to Brad's lessons may have encouraged him to maintain a traditional, "pass-along" instructional model. "Transmitting" mathematical content via a step-by-step, teacher-centered approach to instruction enabled Brad to "control" the information the learners "received." For example, during a lesson which utilized an open-ended, investigative approach to exploring the concept of volume, it appeared that Brad's ultimate intent was to have the learners utilize the formula for volume to determine the number of raisins in a box. The learners attempted various strategies, however none of them appeared to satisfy Brad. He interrupted the learners' discussion to offer his own strategy, "You could measure it, I'll tell you my way of doing it, is I measured in centimeters and I figured 1 cubic centimeter was equal to about 1 raisin, that's, and so you'd calculate the volume of the box." In a critique of this lesson, Brad suggested that, because he could not control the direction of the lesson, the learners failed to do what he had envisioned. As a result, he provided direct instruction on how to solve the problem, "They never really did volume, they just guessed, you know and, I don't know, I kind would have liked them to have measured like the volume of the box and then took a guess on like one raisin equals this." Brad's instructional strategies may also have been influenced by his beliefs about the role of the mathematics teacher. For example, at the onset of the study, Brad explained that the role of the mathematics teacher was to pass on the mathematical skills that students would need in life. He added, "Don't pass along skills that will never be used again, pass along something that will be needed." Brad held to this belief at the conclusion of the study, noting that the role of a mathematics teacher was to "teach mathematics skills that are necessary for future life." However, observation and interview data reveal that, over the course of the semester, Brad's view of teaching and his own instructional practices began to evolve from a transmission, "pass-along" model to a more student-centered, facilitative model. In the following lesson, Brad made an attempt to focus on the underlying mathematical concepts. However, because he directed the learners rather than allowing them to explore on their own, he was not teaching for understanding. Brad began the lesson by giving each student a paper with a number written on it. He explained that it represented their x-value for the equation $y = x$ and asked them to find that spot on the life-size Cartesian axes set up by Melissa. After the students had taken their places, Brad asked them to find their places when the equation is changed to $y = 2x$. Once again, the students found their places on graph:

Brad: Now you notice you're still in a line, what can you tell me about the difference between this line and the line you just did?
Mollie: The slope increased.
Brad: And how was that shown? [As was typical of his style, Brad issued a terse command to the learners and they complied without question or comment]
Mollie: Times 2.
Brad: Now $y = 3x$.
[Once again, the learners find their places on graph.]
Brad: And now what happens?
Mollie: Slope is 3.

Brad: Right, now I'd like to you to do y = -3x [Students find places on graph again.]
Brad: Now, can anybody tell me what's going to happen when we go y = x + 3? [Pause] Is the whole line going to move?
Mollie: The whole line is going to go over.
Brad: Everybody go to the y = x + 3. [Students find places on graph again.] So when you change the number that x is multiplied by, the slope changes, but when you add a number the whole line moves and the slope stays the same, but the position of the line is the same, the same would happen if you did y = x -3, the whole line would shift down, but the slope would still be the same for the line, did everyone get that? Any questions about that? All right, go ahead and sit down. [The learners return to the table and sit down at their places]

Despite the conceptual nature of the above activity, Brad appeared to be focused on moving the learners around the graph and receiving quick and accurate responses to his questions. Rather than encouraging and providing time for all of the learners to participate by articulating their own understandings, Brad permitted one learner to respond to all of his questions. The following excerpt provides an example of Brad's attempts to engage the learners in a mathematical discussion:

Brad: In class on Wednesday, Jamie, it's based on a question that you asked on Wednesday, we talked about, the degrees in an angle, all these adding up, and you were asking about the octagon, if there's a formula for figuring that out, and what you guys are going to do today is discover what that formula is, so to get started, um, all the angles in a triangle added together are 180 degrees? Everybody found that in class, correct, everybody understand that? [While Brad was talking, Wendi was using her pen to cut up soda can]
Wendi: We understand what?
Brad: That the angles in a triangle added up to equal 180.
Wendi: Got it.
Brad: How about a square?
Mollie: They add to 360.
Brad: What I'd like you to do is draw 3 more polygons, not including anything with 3 sides or anything with 4 sides, or where you have 3 angles or 4 angles, something with more, and then add up, each of their angles and figure out, for each shape, how much they are, and you should be able to see a pattern develop based upon the number of sides. [While the learners worked on this task, the preservice teachers circulated to offer individual assistance. Kevin worked alone with Mollie and explained the derivation of formula to her by pointing out the relationship between the number of sides of the polygon and the sum of its interior angles. Brad put a table on the board which showed the number of sides of various polygons and the sum of the interior angle measure of each] What do you notice about these, the difference between them?
Johnny: They're 180 apart. [Brad did not use this opportunity to encourage Johnny to continue talking in order to help him to work through his understandings and discover the formula. Instead, Brad turned to Mollie because he believed she could provide the desired response] Right, so Mollie how would you go about, say you're

coming up with the formula, based on the number of sides, find out the number of degrees, of all the angles added together?
Mollie: What Kevin just taught me?
Kevin: Yeah.
Mollie: I don't know, I can tell you what the equation would be. [Mollie's response indicates that she was only able to reiterate the formula and not the relationships or explanations that underpinned it]
Kevin: We did stuff before that though, before we got to the equation.
Mollie: Finding the, angle number, whatever that was.
Kevin: Well, what's the, what's the, we want to find, the purpose is to find, the relationship between, the angles and how many sides, so how do you do, the relation between the angles and the sides?
Mollie: I have no idea, write down the number, of sides there are and then you add the angle, right?
Kevin: That's what we did, and we noticed what?
Mollie: That, they were all [pause] can't I just tell you what equation I got.
Kevin: No, no.

Rather than allowing the learners time to reflect and construct their own understandings, Melissa and Brad showed them the pattern and the formula. After repeating back the formula to the preservice teachers, the learners continued calculating the number of angles in other polygons. During the break which followed the lesson cited above, Brad remarked to the researcher and the other preservice teachers that he found it difficult to get the learners to think about and discover the relationship between the number of sides of a polygon and the sum of the measures of its interior angles. He recognized that he had difficulty in getting the learners to discover and discuss the formula for calculating the sum of the measures of the interior angles of a polygon. Brad agreed with the assertion that learner reluctance often inhibited discussion of mathematical concepts. However, Brad's own inability to orchestrate meaningful discussion of the mathematical concepts may also have accounted for the lack of discussion. Brad taught several lessons which either began as conceptually-focused or had the potential to be conceptually-focused. Because Brad was unable to utilize learner comments to facilitate discussion of mathematical concepts, the lessons became procedurally-focused. For example, Brad began a lesson on volume by bringing out a variety of solids made with plastic cubes. He then posed a question to the learners which could have been utilized to begin a conceptually-focused discussion:

Brad: Now, we're going to move on to volume, does anybody know, can anybody explain what volume is?
Jamie: Like if you have an almost full glass of water and you put your finger in it and it all spills out the sides.
Brad: What's that the volume of ? [Mollie and Jamie respond in unison, "your finger" Unfortunately, Brad does not utilize the learners' responses to facilitate a conceptually-focused discussion of volume. Instead, he responds to their comments by first asking the learners for a definition of volume but then immediately provides the learners with his own definition of volume] Right, what, like, what's,

does anybody have a definition, for volume? Like it's the amount of space that's taken up, so um, what is, the area of that? Right there. [The learners then spent the rest of the lesson counting cubes and calculating the volume of the solids. At the conclusion of the lesson, Brad asked the learners] Do you have any questions on volume, figuring volume?

Wendi: What exactly is volume?

Brad: Volume, is the total space taken up, like the volume of this room would be the total space in this room, the volume of this figure here, is the total space taken up by this figure based on the cubic units. [Rather than providing the other learners an opportunity to respond to Wendi's questions by allowing them to articulate their own understandings of volume, Brad and Mark gave the learners their definitions of volume]

Mark: And the area would be how many of these little guys it would take to cover the floor and the volume would be how many of these little guys could fill up the whole room.

Brad: Right [As a result, the lesson on volume concluded with the learners restating a procedural or algorithmic definition, rather than with the learners constructing their own conceptual understandings of volume. At this point, Mollie is the only learner to respond to Brad's questions:

Mollie: Ok, so, what you do, is you go, 3 times 4, then 12 times 3?

Brad: Right, and I give you 36, and how many little squares are there?

Mollie: 36

Brad: Does everybody see that? How to figure?

Mollie: Got it.

Brad: Volume, you take the length times the width to find the area of one layer, times the height, so length times width times height, ok?

As previously noted, observations of his teaching revealed that Brad did indeed emphasize traditional methods. During conversations with the researcher and the other preservice teachers, Brad acknowledged that many of his lessons were not conceptually-focused. He alluded to the fact that an inability to integrate mathematical concepts into exploratory activities prevented him from teaching conceptually. Brad also noted that he had difficulty combining mathematical concepts into activities and relating mathematics to "fun-type field trips." Perhaps because of his own traditional focus, Brad was unable to integrate, recognize, or identify the mathematical concepts in exploratory activities:

> We thought we had a lot of good ideas, we had uh, lots of like fun type field trips planned, and so on, and then we had trouble combining mathematical concepts into them, um, trying to get the math related and how we could get the math related and not just have them going on a field trip when we're not doing anything.

At the conclusion of the study, Brad stated that the pre-student teaching experience was "positive" because it enabled him to try out ideas from his education courses in a classroom setting:

> It was a lot better than other classes just cause it was practical experience instead of just sitting in a classroom learning about it, so, that's what really

> was nice about this. The thing I got most out of this was, it was just flat experience, and, I know that my first lesson that I taught was terrible, as compared to the tenth one I taught, and you know, you can take these ideas, and, that you've learned, and put them together, and that was the big thing, this thing was so much better than just writing them down in the classroom, the ideas, we actually got to try them out, and see which ones work or ones not work.

The above comment indicates that the pre-student teaching experience provided Brad with an opportunity to experiment with instructional strategies from his education courses. However, the following transcript of a lesson which took place near the end of the semester reveals that Brad adhered to his closed-ended, procedurally-focused teaching strategy. As in his earlier lesson, Brad attended primarily to Mollie who was able to provide him with the correct responses to his questions:

> Brad: Right, it's the area, or surface area, times, price [writing on board], so that's the price per square foot times the area per square foot, equals total cost to wallpaper that wall, everybody see that? ok?[long pause, silence as Brad writes 2 more problems on the board] ok, this is 2 different walls in your dwelling, and, for this one it costs 70 cents to cover this wall, per square foot, and this wall costs 15 cents per square foot, what is the total cost to cover both of those walls, go ahead and figure it out on your own. [points to a 20' by 80' wall and a 35' by 15' wall, the students work on the problem]
> Mollie: 190.
> Brad: You got 190 for the cost?
> Mollie: Yeah. [the students continue working]
> Brad: Um, did any of you set up a, total cost function, for both of these, like I just set up for the previous one? [shows area 1 and area 2]
> Rianne: Mmhmm.
> Mollie: You can go 8 times 20.
> Brad: Mmm?
> Mollie. 8 times 20, times 70.
> Brad: Right, and that's the cost for this wall.
> Mollie: Yeah.
> Brad: 8 times 20, so what's that? that's area 1.
> Mollie: Yeah.
> Brad: Times, what was that?
> Mollie: 70, times 70.
> Brad: All right.
> Mollie: Plus.
> Brad: Plus.
> Mollie: Area 2.
> Brad: Do you see that? the area of section 1 times the price of section 1, plus the area of section 2, times the price of section 2, and also, it should be noted that if you got 10 walls the total cost function would be like this, just continued out, area 3 times price 3, plus area 4, and on and on, everybody see that? And what we're going to do with your projects now, is uh, you have to find the price, the total cost, of, either carpeting, wallpapering, hardwood flooring, and linoleum floors, your

floors, your walls, for all your, house, you need to come up with the total cost for all of that, and I'll give you some prices of what it cost per square foot, so how would you go about doing that? Johnny?
Johnny: Use that right there. [points to formula on board]

Brad's pre-student teaching experience also appears to have influenced his beliefs about what characterized a good mathematics teacher:

> I think the ability to, uh, relate the math to the students so that they're able to understand it and you have to realize it that not everybody understands it like you and you have to relate to their level so that they can grasp what you're trying to teach.

Kevin

Kevin clearly enjoyed the personal contact with the learners and appeared to have forged the strongest personal relationships with the students. Kevin worked at building a positive relationship with the learners. When he arrived in class he would engage the learners in small talk. He had the ability to pick up on a comment a learner had made during a previous class and use that to begin a conversation or lesson. Kevin identified the strengths of his teaching as, "Oh, I think, I think I understand where each student is pretty much at, I mean I can, I have an idea of who I'm talking to when I say what, and I'll try to, say different things to address different people." Kevin's interactions with the learners may have been influenced by his beliefs about teaching mathematics. For example, Kevin identified "caring" as an important characteristics of a good mathematics teacher:

> Caring, I think that's probably most important, a characteristic of a good teacher, for a math teacher it's uh, (pause) well, they should know math as well that might help, but I think, I think if they care about their students, they care about their students' learning, I think they're going to be a good math teacher.

Kevin had fairly high self-concept in both mathematics and teaching and clearly enjoyed the subject. His high self-concept appeared to remain throughout the semester. During the planning sessions and in discussions with the researcher, Kevin revealed that he had a high self-concept in high school mathematics, "Oh yeah, I'm fine with pretty much anything that's taught in high school." Kevin admitted that he tended to "stumble sometimes" over the mathematics he encountered at the university. However, this did not trouble Kevin because, although he enjoyed his college mathematics classes, he didn't see the relevance of some of the university's required mathematics courses to the courses he would be teaching as a high school science and mathematics teacher. Kevin returned to this theme of "relevant" mathematics when describing a metaphor for mathematics:

> Math is like, it is for me, it's like looking over the horizon, and if you've never been beyond that horizon you can see it and it makes sense and you've been there before and then beyond that horizon it's abstract, you don't know what's out there, it's weird and, and if you have everything you need right here, sometimes you're like, I don't really care what's over there and that's kind of what it's like for me because, I, there's theoretical math, higher math, I personally don't care (laughs).

Kevin maintained that certain people possessed talent in mathematics. However, his comments during an initial interview suggest that he did not believe that a lack of talent precluded a learner from succeeding in mathematics:

> I, it's nice to have talent and you could be better at it if you have talent but like at the same time I think that hard work will get you to the same place that most people can get on talent I mean when you start talking upper level and abstract math, I think that's different like in high school sports, if you work hard you can pretty much play, I mean if you're relatively gifted but then you get to the college level and people work hard but if you don't have the ability, you don't have the gift.

While Kevin described his experiences with high school mathematics as positive, he believed that, because he found the classes fairly easy, he could have been placed into higher level courses. He admitted that his failure to take precalculus and calculus in high school made his college mathematics courses more difficult. Kevin's enjoyment of and success in high school mathematics did not seem to be adversely impacted by poor teaching:

> Well, the teaching, well my sophomore year in geometry, I couldn't stand our teacher, it was his last year teaching and, he, it seemed like he just didn't care, he used to get up there, do an example and say "go ahead" that was all, and so that was, I did just fine in the class, it was just because I couldn't work from the book, I had to figure stuff out on my own but, but it was, it was rather difficult just from the fact that he didn't teach but other than that I had some really good teachers just, what math, the math that I did came pretty easy.

Kevin did not rely on written lesson plans and appeared to be very comfortable with extemporaneous and loosely planned lessons. At the conclusion of the study, Kevin noted that, as a result of the experience, "I think that I'm a lot more confident as a teacher, definitely." Kevin believed that the most important reason for teaching mathematics was to improve the self-esteem of learners, "To make a kid feel good about themselves because math, math is often, it's a difficult subject and I can see a big esteem difference between people that think they can do it and people that don't think they can do it." This belief was evident in observations of Kevin's student-centered instructional practices and his interactions with the learners. In addition to getting to know each of the learners on a personal level, Kevin appeared to be genuinely sincere in the praise, encouragement, and positive feedback that he gave to the learners. It appears that the personal bonds that developed between Kevin and the learners resulted in a change in his perceptions of students on the margin:

> Before, I always pictured these kind of students, the at-risk and disengaged and ones that I recall from my past as being [pause] oh, harder to get along with, almost to the point of being outcasts, but I didn't feel like, these students weren't outcasts at all, it was just, some things in their life that have happened that have put them in a certain situation, um, I think it, just like anything, you can't really stereotype these students cause each one is certainly different.

Kevin's belief, which connected success in mathematics to improved self-esteem, appears to have been shaped, in part, by his high school experiences. He explained that his high school was rated one of the top three in the state academically. Moreover, the high school's learners came from homes with the state's highest household incomes. This combination of demographic factors resulted in 98% of the learners attending college after graduation. In addition, a large number of learners had parents who were executives at the headquarters of a major technology corporation. Kevin noted this had a tremendous influence on the character of the school, resulting in a high status being conferred upon learners who enrolled in honors computer science, mathematics, and science courses. Observations of Kevin in the classroom and discussions which took place during the planning sessions and outside of class suggest that Kevin genuinely cared about the learners. While his beliefs about the importance of caring were essentially unchanged at the conclusion of the study, Kevin's comments indicate that the pre-student teaching experience had convinced him of the importance of utilizing multiple methods, perspectives, and explanations into his instructional practice:

> I think a good math teacher [pause] is patient, and a good math teacher can see things from different ways, um, it is kind of funny that, all the math teachers are the ones that got it relatively, pretty easy the first time through, and, and, I've been at the point a couple of times, like tutoring, I just cannot understand why these people don't understand what I'm explaining, and I think a really good math teacher has the ability to say, ok, this isn't working, how can I change what I'm saying, how can I form [present] this from a completely different angle to try and help these people understand it.

It appears that Kevin's high mathematics self-concept provided him with the confidence to engage in spontaneous, open-ended instructional strategies. If he tried a new idea and it did not work out as planned, he did not lose his self-confidence or appear shaken. Rather, as the following transcript of a lesson on scale reveals, when Kevin faltered, he would backtrack, move on to something else, or ask the other preservice teachers for assistance:

> Kevin: Careful, let's do this one, do this one, we're going to draw that wall, we're going to draw this wall right here, ok, how long should I draw this line, if we're

drawing the scale, if I write the scale, you've got 1 inch, to 22.2 inches or you could write 1 inch over 22.2, does it matter which one's on top or the bottom? [turns and asks the other preservice teachers who are present]
Toni: It depends on whether you're going up or down scale.
Kevin: Yeah, yeah. [returns to lesson and leads the learners through the process of scaling down a wall] Ok, that's our scale, right there, that's our scale for these 2 walls, that's our scale for these 2 walls, and we could finish drawing the box, if we want to, but, I think we went at it a real complicated way, there's a lot of stuff to remember if you do it that way.
Wendi: That was fun.
Kevin: You think so, could you, redo it, if I erased it, and gave you a problem? [both Jamie and Wendi respond, "no"]
Kevin: Exactly, um, the way I would have done it personally, I would have said, ok, I have 22, 22 feet, you can leave it in different units as long as you write those units, I would have said, let's see, I want to go to, oh, I want it to go to 12 inches, 12 inches is to, 22 feet, that right there is a scale [pause] right? Is that a scale? Can I reduce that at all? [Kevin demonstrates his method] right, right, this is, this is a real good way to do it, when you don't have to do all that, if you don't want to.

Early in the semester, many of Kevin's lessons were procedurally-focused and predominantly teacher-centered. For example, rather than focusing on the meaning of and the relationships between fractions and decimals, Kevin walked the learners through a series of problems using a step-by-step approach:

Kevin: Let's start with this one... [pause] somebody read this for me.
Wendi: 3 divided by
Kevin: How else can you say it? [Wendi and Mollie both respond "3 divided by 5"]
Kevin: Ok, which can be written 3 divided by 5. How else can that be written? 3 divided by 5? yeah? Ok.
Jamie: No.
Mollie: You can't?
Kevin: Isn't that right?
Rianne: The line has to go on the outside.
Kevin: See, it's 3, no Rianne, you're right, 3 is divided by 5, ok? And now how would you go about solving that, Rianne?
Rianne: Put a point there.
Kevin: So I need to put my point here?
Rianne: Yeah.
Kevin: If I put it here I also have to bring it up here, right? Then I could put a zero here, then what do I do?
Rianne: Put a 6.
Kevin: 6, and that gives us a remainder of 0, so we're done, so point 6, 3/5 equals point 6, in decimals. Now, let's move to a number that looks a little more complicated. How do we do this? Wendi? [the lesson continued in the same manner]

However, as the semester progressed, Kevin's lessons became more student-centered. He also began to integrate open-ended activities based upon real-life examples into his lessons. For example, later in the semester, Kevin attempted to engage the learners in a conceptually-focused discussion of ratio and scale by utilizing real-life examples, including a scaled map of Inland State University's campus:

> Kevin: What's what's the easiest part to measure, and what measurement do you know?
> Wendi: Uh, the field, the field [referring to the football field].
> Kevin: How long is it?
> Wendi: 100 yards.
> Kevin: Good, it's 100 yards.
> Wendi: So it's from here to here.
> Kevin: That's not including the bleachers, so what would you measure on this? Measure just the field?
> Mollie: It's about 1 ½ inches?
> Kevin: Yeah.
> Wendi: You'd measure the field.
> Mollie: And the bleachers go along with it.
> Jamie: Ok, they did not, no it's not 1 ½ inches, they didn't use
> Mollie: It's about 1 ½ inches, yeah, it is exactly 1 ½ inches.
> Jamie: Oh.
> Mark: So then take it down to this scale.
> Kevin: And what is 1 ½ inches?
> Mollie: Umm, 300 whatever.
> Wendi: 300 yards.
> Mollie: So, what, on that scale?
> Kevin: So we know what, we know, 1 ½ inches, we know that 1 ½ inches is to what?
> Mollie: 300 yards, oh no, wait, 1 ½ inches is going to be, 100 yards.
> Kevin: 100 yards.
> Mollie: And then 400 yards is to 300, whatever
> Kevin then poses a question which requires the learners to think deeply about proportional reasoning:
> Kevin: So, it's 1 ½ inches to, so, what if, what if we want to find out what 1 inch is, who can tell me how to do that?
> Mollie: Oh it's 200.

Kevin's critique of the university's education professors suggests that he did not model his own instructional practices on the teaching he observed in his education classes:

> Those [education] professors, I don't know if they're from, just because they're from the education field, I would assume that they would use the practices they preach but it's always, it's just a straight lecture, nothing else, and you would think that if they're telling you about these practices they might implement it a bit but I've never experienced that and along with that I've had some of the most boring teachers, I think, have been in education, they just talk and they say so much and they say so much and so little at the same time.

Kevin's use of innovative instructional strategies may have been related, in part, to his perceptions of marginalized learners. Kevin believed that at-risk learners were unmotivated students and, as such, "You have to provide them a hook, a why, whatever's important to them, that's the big thing." Observations of Kevin's teaching revealed that he did indeed try to provide the learners with a "hook" at the beginning of each lesson. For example, Kevin began a lesson on angle measurement by jumping in circles and challenging the learners to guess how many degrees he had rotated. Kevin believed that disconnected learners had different motivations to learn mathematics than their mainstream peers. His explanation was based upon his participation in the pre-student teaching experience and his interactions with the learners:

> I think, well at least from what I saw, graduating from high school is a, is an adequate enough goal for a lot of them, and, so I think you have to change your teaching philosophy, where I, the high school I went to, if, I think, 97% of the students went on to college, and, I think that's a lot different from what they get at the alternative high school, and so, our, a lot of our motivation, our teachers used on us was, you know, you gotta get good grades, if you're going get into college, and you need to learn this so you can write good essays in college so you can take physics in college and so you can do this in college, well, when your students are just trying to graduate, from high school, you have to motivate those students a lot different, and I think we tried a lot to motivate them by saying look, this is real life, it's real, they know it's real, I think that's it, a motivator for them.

At the conclusion of the study, Kevin indicated that his experiences teaching at the alternative high school reinforced his prior beliefs about the role of self-confidence in learning mathematics:

> I think if anything, I always thought that math had a lot to do with self-confidence, not so much, I guess the question was, teaching math, but it, I guess, my response to that would be that we're teaching self-confidence, because the students, at the beginning of the semester, they saw, they just didn't think they could do it, but then at the end of the semester they'd say, oh, I can't do that, I don't know how to do that, but then, they could do it, so I think, teaching self-confidence is what I learned about teaching math.

Kevin's beliefs about mathematics may also have played a role in the evolution of his instructional strategies from procedurally-focused to conceptually-focused. Of the five preservice teachers who participated in this study, Kevin was the only teacher who explicitly articulated how his beliefs about mathematics were changed as a result of the pre-student teaching experience. He remarked that his ability to do mathematics was not sufficient to enable him to teach mathematics. Kevin's comments suggest that he rejected the view of mathematics as a body of facts to be transmitted directly to learners. Observations of Kevin's teaching revealed that his belief was reflected in his instructional practices which utilized multiple methods and encouraged learners to explore and discuss the mathematical content areas. Kevin's instructional strategies may also have been influenced by his beliefs about the role of the mathematics teacher. Kevin described the role of a mathematics teacher as being a "coach". Observations of Kevin in the classroom revealed that he could indeed be characterized as a coach. Although he often allowed the learners to discover on their own, he was always willing to provide guidance and alleviate confusion:

> You want the kids to learn on their own as much as they can, you want make sense for them and not just memorizing steps out of a book, you want it to make sense to them, and, and, almost be a part of them, almost to where it makes sense to them, but then it's also your job to push them, it's your job when they're confused to help them, it's your job to give them direction and, and what they're going to learn, it's your job to lay that out.

During his lessons, Kevin often discussed mathematical concepts. Although he demonstrated conceptual understandings of many of the mathematical topics, he appeared to be unable or unwilling to engage the learners in these conceptual discussions. Instead, Kevin talked about the concepts while the learners listened. After viewing a video clip of his lesson on ratio and scale, Kevin critiqued his instructional practices. He referred to this same lesson on several other occasions during the study. Kevin explained that while he was trying to teach the concepts using an exploratory approach, the learner was expecting the formulas delivered via a step-by-step approach:

> He [Johnny] was at the point of frustration, where he was saying I don't understand any of it and, and, it doesn't help that there was a couple of different teachers teaching it, I think a couple, in different ways, and so, I don't think, I think that makes a big difference for a lot of the students that are trying to, memorize what to do, I guess, and, I mean they're just trying to memorize something, and they're like ok, what's the first step, what's the second step, and I don't think Johnny was getting that, and if you look at Johnny, that's what Johnny wants, is, to do, ok, what do I do first, what do I do second, and how do I get the answer, yeah, I think, was what I did from that point is I tried to get Johnny to be able to do it as opposed to going back around to the concepts and I think that, that's what was important to him, and I think that then I would have gone back and talked about the concept of scale again, in simple terms, and then say, ok, that's what scale is, and then give him an easy example that he could do, just that he could say, "oh, I could do that", and I know that's what, he wants to be able to say, is that, "oh, I could do that", and then he's got that confidence.

The following transcript of a class Kevin taught on angle measurement is typical of his attempts to integrate real life examples into a conceptually-focused, student-centered lesson. In an effort to engage the learners and encourage discussion, Kevin began the lesson by asking students if they had ever been snowboarding. Before launching into the lesson, he gave Jamie time to tell a story about a rollerblading mishap.

> Kevin: Ok, you guys, I see you guys are at least familiar with these kinds of things, right? You guys ever think of kick flips you might do, kick flips, skateboarding, what kind of tricks might you do?
> Mollie: Like 360 and 180 and those things.
> Kevin: One of my friends, whose my best friend, loves to go boarding, and uh, he can do 720s, you guys know what that is?
> Mollie: You spin around, er, 2 whole circles.
> Kevin: Right, so we're talking about different tricks, come up with a different ones, we talked about the 180, the 360, 720s, do we know what all that stuff means? [At this point in the lesson, Kevin encourages the learners to relate snowboarding to mathematics:]
> Jamie: Circles?
> Kevin: Circles.
> Jamie: Degrees of circles. [In an attempt to encourage the learners to think about the underlying concepts, Kevin asks open-ended questions]
> Kevin: Degrees of circles, so what do you guys know about degrees of circles?
> Mollie: Uh, full circle is 360. [Rather than utilizing inert diagrams to demonstrate degrees, Kevin uses his body as a dynamic example]
> Kevin: So start, basically a reference point, right, you were facing this way, you jump around 360 degrees, you come back to the same place, and when it's 0, you don't jump at all right? Now what if you do a 180? If I'm facing this way, what way would I face when I do a 180?
> Wendi: You're going to be the opposite way. [Kevin recognizes the learners' comments and utilizes them to clarify explanations and definitions and transition to the next segment of the lesson]
> Kevin: So it would just be like a straight line, right? It's the opposite direction, good, so you guys, are at least familiar with different angles, and now, what do we do if we don't exactly get all the way around, I mean how, how far do we turn then? And how are we going to estimate those? Like, what if, what if, what if you can't jump very high and you can only turn that far? [Mollie and Wendi both respond "90 degrees"]
> Kevin: Good! What if I jump up, and, I can't quite make it around and I go that far? [Kevin is standing up jumping and turning] [Throughout the lesson, Kevin encourages the learners to articulate their observations]
> Mollie: 180 plus 90. [Kevin utilized the learners' comments to clarify and transition to the next segment of the lesson]
> Kevin: Uh huh, which is, 270. Ok, now, in math, most of the time, mmm, especially in geometry, I suppose, most of the time you're dealing with angles that are less than 180 degrees, because you can say this, you can say, how far do I go, this way, how many degrees? [Learners respond, "180." Then, as Kevin jumps and turns as learners identify the degrees Kevin jumped in relation to where he started.] I guess, seems like you guys are pretty familiar with angles, right?
> Johnny: Kind of, not really, I never took this. [In an effort to encourage the learners to connect the concepts in the current discussion to other mathematical topics, Kevin prods the learners to discuss their prior experiences with angles]

Kevin: What are your experiences with angles? Do you remember anything? Triangles?

Johnny: I just know like the 90 degree angle.

Kevin: Basic stuff right, so, I have an idea. [Kevin urges the learners to jump. When they decline, Brad, the other preservice teacher, jumps as the learners guess the degrees he jumped]

Mollie: More than 90 less than 180.

Kevin: Right, right, so he's somewhere between this angle and this angle, so you could also say that's, 180, 270, ok, now it gets kind of complicated when you think about the whole circle, that's why in geometry, you normally go with?

Jamie: The half right? [At this point Kevin introduces the protractor and students begin to measure angles.. Kevin then goes to the life-size Cartesian axes and utilizes it as another way to demonstrate angle measure]

Kevin: Right, 45, and this is 45 but what else could we say? This angle, and this angle, together, they're both 90 degrees, from here over to here, is 90 degrees, you see, all an angle is, it's a way to relate 2 lines, or a way to relate, 2 directions, right? Just like we were talking about these different skiing jumps or something like that, all it is, is a way to relate, which way you're facing before you jump with which way you're facing after, that's what an angle is, ok? You guys see that?

Mollie: Yeah.

Kevin: Yeah, ok, that makes sense to everybody right? [Students answer, "Right!"]

Wendi: So far.

Kevin: Ok.

Wendi: I get it. [Kevin continues to encourage the learners to participate in the activities and articulate their observations:]

Kevin: Ok, so now, now we know what an angle means in a relation between 2 lines, right? We know, we know that, if we have a line like that, and that's some angle, let's estimate that.

Wendi: It could be close to 45, I mean, close to, yeah close to 45.

Kevin: We'll say 45, ok?

Mollie: All right.

Kevin: So here's a 45 degree angle, what's that?

Wendi: 90, no, that's?

Kevin: Any of you seen the symbol for 90 degree angle, it's like a little square you put in the corner here, that means 90 degrees, so we know that this line's greater than that.

Wendi: 130, 135.

Kevin: 135, good, so this angle's 135, what happens if you put a line right there? Then we have to figure what that angle is, ok? That's when it gets tricky, ok? [whiteboard falls down, laughter] Ok, now we're going to do angles in motion, ok, I have triangles for everybody ok? [gives the learners paper triangles to work with in the next segment]

The previous segment illustrates several features that characterized Kevin's instructional practices. Among these were Kevin's his ability to utilize multiple examples, including examples drawn from real-life, to illustrate mathematical concepts. Throughout the lesson, Kevin utilized several strategies to encourage learner discussion. These included encouraging the learners to articulate their observations, discuss the mathematical concepts in relation to real-life examples, and explain their prior experiences with the concepts. Kevin also utilized learners' comments as a springboard for his own explanations, to present definitions, and to transition into the next phase of his lessons. Kevin made attempts to connect the concepts in the current discussion to other mathematical topics. He accomplished this by encouraging the learners to discuss their prior experiences with angles. At an end-of-the-semester, Kevin offered his thoughts on the significance of the pre-student teaching experience:

> I think that talking about students with problems was one of the biggest lessons for me this semester, you're dealing students that had, a lot of personal problems, and, and although you don't want the personal problems necessarily to be an issue in the class, still they're interwoven with the person, and you can't just toss them out, as teachers, this is, this is what I've really come to feel this semester, is that, you're not so much just teaching the content, you're really teaching the students, you're really teaching them, you're really being a role model to them, and, and also, you have to be understanding of their situations, and, you need to be interested in students lives, I think that's extremely important.

At the final interview, Kevin was once again asked to identify what he believed was the most important reason for teaching mathematics. Referring to his experiences at the alternative high school, Kevin's response of "life skills" referred to logic, reasoning, and analytical thinking. Kevin's belief that "life skills" mathematics consisted of logic and reasoning rather than arithmetic skills was reflected in Kevin's use of investigative teaching strategies.

> Well it's, just like in our class, we saw different ways that math is used in real life, in the class we taught, yeah, and, and I think that most of the stuff that students learn in the class won't apply in real life, but the logic used in math is, and it's, it's a skill, the logic used in math is, is a skill that, that is really valuable, it's, it's, the logical skill, I mean, is it this, or is it this? it's a real analytical viewpoint of life, either it's right or it's wrong, and, and then multiple intelligences, I suppose.

Possibly because of his pre-student teaching experiences, at the conclusion of the study, Kevin discussed the role of the mathematics teacher in terms of his personal vision. He once again noted the importance of building learner confidence. Kevin also stated that he was "not so wrapped up into necessarily teaching math as I am to teaching students.":

> I would see my role as, as a [pause] a teacher, I mean it's not, I'm not so wrapped up into necessarily teaching math as I am to teaching students, um, and, and of course, I would want them to learn math, because I think it's important, especially, not just, to achieve in my class or achieve in future classes, but just to have those skills, but, um, I think it's more important to, give those students confidence, um, let them [pause] I guess just be an

> example as well, I think that's one of the most important parts of being a teacher.

Toni

Of the five preservice teachers who participated in this study, Toni appeared to have the lowest self-concept in both mathematics and teaching. Despite her low mathematics self-concept, Toni indicated that she liked mathematics. Her metaphor for mathematics revealed that, even though she struggled with the underlying concepts of mathematics, she enjoyed the subject, "Math is just this giant banquet, you help yourself to, you know, there's all kinds of things you pick and choose from and they're all delightful, most of them anyway." Although Toni cited the perceived usefulness of mathematics as a motivator for learners, she noted that, "mathematics for its own sake is enough for me." During her initial interview, Toni spoke of an early and life-long interest in numbers:

> I've just always had an interest in numbers, ever since I can remember I've always had some fascination with numbers. I was always counting, looking for patterns, I would see patterns with music, I would close my eyes and see geometric patterns with the music. So with everything it's always been there, since earliest memories, it's been second nature, I never thought about it as being different, looking for patterns. Oh, I kept tablets, you know those Big Chief tablets, I kept a tablet and instead of when you're bored you go write in your diary, how some people will keep a diary, well, I kept a tablet and instead of you know, I didn't work anything, I would draw columns and I would start with one and I just kept tablets of numbers and I started that I could just go 0 1 to 9, fill up that column and then move to the next one, I was playing with place values, when you look back on it, it kind of looks real stupid, well (laughs), numbers.

Toni's love of numbers appears to have encouraged her to persist in mathematics despite her negative experiences with school mathematics. She revealed that she had "problems with testing" and felt it was unfortunate that much of mathematics learning was measured by traditional forms of assessment, including written tests. As noted earlier, Toni did indicate that many of her negative experiences with university mathematics courses were mediated by positive interactions with women mathematicians. At the onset of the study, Toni indicated that the most important reason for teaching mathematics was to mentor and motivate learners and provide examples of "ordinary" people enjoying mathematics. Subsequent discussions with Toni revealed that this belief appears to have been shaped, in part, by her background and negative experiences with school mathematics:

> For me personally, I think the number one reason that's important for me to do that is because I think I never had the mentors and I think it's important, like I said in the very beginning that it's important for both boys and girls to see people doing math, that like to do math, and not necessarily being really

> good at it, what some people might term genius in math, but someone who enjoys it, you don't have to be Einstein to enjoy math, you don't and also someone, a woman being someone who enjoys math, those are the key motivators for me.

Toni initially appeared uncomfortable in front of the classroom. Observations and interviews revealed that this discomfort was a result of both her teaching inexperience and her weak content knowledge. Toni was so uncertain of her teaching abilities that, halfway through the semester, she requested that she not be shown a video clip of her teaching. Toni explained that seeing her own "weaknesses" might further erode her confidence. However, as the semester progressed, Toni appeared to gain in confidence. While her low self-concept was evident for much of the semester, by the conclusion of the study, Toni's confidence in her teaching abilities appeared to have increased. According to Toni, it was not until the last class that she comfortable, essentially because she had more confidence in herself and was "handling things." At the conclusion of the study, she affirmed that her confidence had grown as a result of the experience:

> It's kind of helped me fine tune, the kinds of things that I need to be aware of, I, uh, I really felt like the last class that I was actually teaching was going, really nicely, I mean it wasn't perfect, it wasn't absolutely the way I would like it, but, but it was, it went, well, with the probability, and pi, and whatnot, I mean there were some, I, I, uh, that, I kind of got thrown off what I was going to do for that, but, uh, in a similar way, to uh, I had, gotten thrown off in one class at the very beginning of the semester, but this time it was, oh, it, it, it, did, I just was able to, cause of my own confidence in myself, and, even though I didn't know the material as well as, I wanted to, I, I, I was, it was good, it just felt good, it felt like I was, you know, handling things and, uh, it was good.

Throughout the semester, Toni alluded to the fact that, because of her own educational experiences, she was often marginalized and disconnected and had been considered an "at-risk" learner. She began a discussion of students on the margin learners by stating she had a problem with labeling students. For example, Toni believed that the "at-risk" label, "puts it all on the student." Noting that there are probably at-risk learners in every classroom, Toni identified a variety of factors that she believed marginalized students:

> There are a lot of reasons for a student getting to that point - it can be anyone who isn't encouraged, who doesn't see a reason to learn math, isn't motivated - and a good math teacher can help a student become motivated and find enjoyment in math. These are the students you're going to lose, that will drop out - and they don't always come back.

Toni indicated that reaching students on the margin required a mathematics teacher who could provide learners with a variety of strategies for doing mathematics. This belief was evident in Toni's efforts to implement multiple methods of instruction in her own mathematics lessons:

> Hmm, uhm, it takes a teacher who likes math, enjoys math - and who understands the different ways that students learn and understand math, and getting away from the idea that a student is stupid or just doesn't have what it takes if they don't follow a certain way of teaching math - they may need something different, that there are different ways of learning and doing math.

Toni indicated that pre-student teaching experience did little to change her beliefs about at-risk learners. This is consistent with the fact that Toni believed that, by some definitions, she herself was a student on the margins and could have been labeled as an at-risk learner. Toni's comment suggests that, unlike the other preservice teachers, her expectations of the learners at the alternative high school were not contradicted by her experience, "No, not really, I don't think, [pause] I don't think it's really changed it that much, going in I had some, expectations, but, [pause] not unreasonable ones, and none that were contradicted really either." While there was no apparent change in Toni's perceptions about these learners, at the conclusion of the semester, her perceptions appear to have been refined, in part, by her interactions with the learners:

> Ooooh, [pause] I think it's a little bit different in that the intensity with which you have to, have to, be aware is different, uh, I mean you still have to be aware of other students who aren't marginalized, and what they're doing, and what's going on with them, but with these kids you, you really need to be in tuned to what's going on, and what will affect them, and what you say, and, and things like that, and, what, you know, what's affecting their work in class, that kind of thing.

Perhaps because of her focus on self-confidence and her own negative experiences with traditional mathematics instruction, Toni was extremely willing to attempt to teach for understanding. In addition to her willingness to experiment with open-ended, investigative activities, Toni attempted to design student-centered learning environments in which the learners could explore the concepts underlying the mathematical topics. For example, in order to teach a lesson on ratio, scale, and proportion, Toni arrived in class with a large carpet bag that was overflowing with a variety of miniature dollhouse tools, life-size tools, measuring tapes, rulers, pencils, and magazines. At first, each learner selected a miniature tool, paired it with a life-size tool and then attempted to determine the scale between the tools. The learners were next given schematics of birdhouses and asked to determine the scale. Perhaps because of the novelty of the materials, all of the learners were extremely attentive and involved in the lesson:

> Toni: Let's try and figure out the scale, let's figure out the scale they used, what would we have to do to figure out the scale of these?

Jamie: They use uh, equations kind of too, like part, name, size and pieces.
Rianne: How you find the scale?
Toni: Yeah, what do you need to know in order to find the scale? What did we need?
Rianne: How tall it is?
Toni: Right, what do we need to know when we
Jamie: You have to find the top, side.
Rianne: How tall it was, how long it was.
Toni: How long it was.
Rianne: The length.
Toni: How long the picture is? Or how long, the actual size is?
Rianne: Yeah.
Toni: Ok, so we need to know actual size.
Johnny: You need the dimensions.
Rianne: Yeah.
Toni: Well, of course you need the dimensions, I'm assuming that you said the actual dimensions, but in order to figure out the ratio, did we [pause] just do it from this? [pause] What else do we have? Jamie?
Jamie: We have, the ratio, the small, compare it.
Rianne: The small one, compare it to.
Toni: With the small one, compare it to, good, because ratio is a comparison, like we said earlier, so we need the small one, so what's the comparable thing here, if we know the actual size of our bird cage, what else do we need to?
Jamie: You need to figure out the small size.
Toni: Small size, I'll say reduced size.

Toni designed a review lesson which she believed would assist each learner in developing their understanding of concepts which had proved difficult for them. It appears that Toni's beliefs about teaching mathematics influenced her strategy:

> I think it's a result of basic desire, uhm, how successful you are in math is for me, depends on how much you want to be successful and want to do it and it is dependent on your experience with math, if you can't find someone who understands how you learn math and how you relate to it, they're not going to be able to teach you in a way that won't frustrate you, that's what I think, you have to be able to.

While Toni's efforts at teaching for understanding were not always successful, she persevered in trying to provide learners with student-centered activities and conceptual explanations. She was also persistent in attempting to engage the learners in discussions in order to elicit conceptual explanations from the learners and encourage them to articulate their understandings. It appears that, unlike Brad, Toni was extremely willing to integrate many of the ideas from her education courses into her mathematics lessons. In the following excerpt, Toni was attempting to teach the concept of area for understanding. In addition to engaging all of the learners, Toni appeared to have some measure of success in eliciting learner explanations:

Toni: If you have this, um, same, dimensions, 3 by 5 [points to a 3 x 5 rectangle], but, somebody comes along and goes, shifts it all that way, right [draws a 3 x 5 parallelogram] [pause] now we've got um, we've still got 3 inches here, and you've still got 5 inches here, and you've still got 3 inches here, and then we've got 5 inches here, are we going to have the same area in here? Is that area going to be 15?
Rianne: No.
Toni: Are you sure? [pause]
Johnny: How could it change it?
Toni: That's a good question, when I slant it what happens, what happens here? [At this point all of the learners were clearly engaged in the activity.]
Johnny: The angles change.
Toni: Here's my 3 by 5, and here I have, a 90 degree angle, and here I have a less than 90 degree angle.
Jamie: It's still all going to be 360.
Toni: We'll just name that theta and we'll say it's less than 90 degrees.
Jamie: It's still going to be, it's closer, these here are bigger than 90 degrees, these 2 here, are less than, so it's still going to be 360.
Toni: If this is 3 inches right here, the same as this is 3 inches, how far is it from here to here? On the square, on the rectangle? What's the distance?
Jamie: 3 inches.
Toni: Ok, what's the distance here [on the parallelogram]?
Jamie: 3.
Toni: Are you sure about that? If this is 3 inches and it's less than 90 degrees?
Rianne: Why is it 3 and 5?
Toni: Because I kept it the same, I want that to be 3 inches, ok, I want that line to be 3 inches long, this is still 3 inches long.
Wendi: But it changes because it's slanting, it changes.
Toni: That's what I'm asking.
Rianne: When you slant it, it gets bigger? [The learners suggest that "slanting" the rectangle would change the length.]
Toni: That's what I'm asking, that's what I'm asking, this is still a 90 degree angle here, [pause] ok, it's the same angle.
Johnny: So it wouldn't be the same area.
Jamie: Still, all those are going to equal 360.
Toni: If I?
Jamie: Those 2 are obtuse, and these ones are acute, but they're still, all going to equal 360. [Jamie recognizes that the internal angle measures for each 4-sided polygon will equal 360]
Toni: All right, what's, what's?
Johnny: I'm not sure they're the same size though.
Toni: Ok, well let's take a look at this, do you think it would be the same?
Jamie: But I mean the middle one, I mean if you took it and slanted it, it wouldn't be 3 inches any more, the corners [pause] so I guess the middle wouldn't be 3 inches, the area wouldn't be, the lines?
Toni: The lines can still change, what if my, like Jamie just said, what if this was still 3 inches, this was still 5 inches, so it's, all these were the same, am I getting closer to this line?
Jamie: That's changing.

Toni: So this, this number here is not going to be 3 inches.
Jamie: They all have to be 3's and 5's.
Johnny: Just cause the sides are that doesn't mean the area's going to be, the multiplication of the 2, right?
Toni: I think you're on to something, because if I ended up, I could end up with collapsing that all the way to the ground and I'd still have a 3 inch side and a 5 inch side laying on top of the 3 inch side and the 5 inch side but my area would be?
Rianne: Nothing.

In the initial interview, which took place prior to beginning the pre-student teaching experience, Toni spoke in the third person as she described the role of the mathematics teacher. At the final interview, Toni was able to put her response in personal terms, indicating how she would define her own role as a teacher of mathematics, "I'd envision myself as an application, I guess, someone who's going to [pause] help them figure out how to do it, you know, help them figure out one that they can do it, and two that, they can do more than they think they can." Toni often appeared uneasy when teaching. This uneasiness may have resulted from her recognizing that she had insufficient content knowledge. Toni's belief that mathematics teachers were supposed to "know everything" may have served to intensify her apprehension, "I still have [a belief] from the old school, that when you're a math teacher, you're supposed to know everything [laughs, hitting hands together] you're supposed to know EVERYTHING!" Although Toni identified a good mathematics teacher as one who was "able to listen", she was not always able to listen to the learners. It appears that her own weak conceptual understandings inhibited her ability to listen and respond to learner comments:

> Oh boy, somebody who listens for sure, um, and someone who um, [pause] knows how to find out, what the students need to know, you know, whether it's within their expertise, or outside their expertise, you need to find out what, what they need to know, and willing to not, not get upset when they know more than you do.

Unfortunately, because of her own weak conceptual understandings and resistance on the part of the learners, Toni's efforts at listening to and understanding learners were not always successful. For example, in an effort to extend the learners' understanding of the concept of slope, Toni brought in magazine photos of birdhouses and asked learners to calculate the pitch of the birdhouse roofs. She made an effort to have the learners discover the relationships and to connect those to the notion of "rise over run." However, the two learners working on this task were confused about the terms "pitch" and "relationship":

Toni: [to Wendi] What I want you to find the relationship between the distance from this wall and where that peak goes and the height, think about it [pause] before you ask a question, I want you to find the difference between the, the relationship between, the height from here, and it's distance to here, ok? [to Mollie] Mollie, [pause] I want you to do that same one, and you can use, any of the structures, on these pages here, ok? [pause]

Mollie: I don't understand what you said.
Toni: Well, think about it for a minute.
Wendi: It's confusing me.
Toni: You have a height here and you have, this one doesn't have a good wall.
Wendi: Shucks.
Toni: But, I, I want to know the distance, if you drew a line straight down the middle, I want to know the distance from the outer edge and the height, and I want a relationship.
Mollie: What's a relationship?
Toni: Well, we've been discussing ratios, and scale, so find the relationship between those two [pause] numbers [pause] in context with what we've been discussing.
Mollie: So I just have to find out, what?
Toni: Well, think about it, think about what you have to find out.

In his reflection on this lesson, Brad stated, "The students seemed confused with the word 'relationship' between two numbers. After they figured out what they were doing they discussed what they found. They seemed uncomfortable in describing what they had done." Because of the learners' own minimal understandings, and Toni's inability to clarify the definitions for them, the learners were unable to realize the connection between the birdhouse roofs and the previous day's lesson on graphing and slope. At the conclusion of the study, Toni indicated that her teaching experiences at the alternative high school had not changed her beliefs about teaching mathematics. However, her responses did suggest that she was able to clarify and expand upon her prior beliefs. Toni explained that the overall experience helped her to recognize "the kinds of things I need to be aware of" and provided her with an opportunity to "help me fine tune how I want to do it, more, but not changed it really." Toni made a statement which reveals another way in which the experience had changed her beliefs about teaching, "The important thing to remember about lesson plans, at least what I got from this, is that, it is a plan, ok, and plans change, ok, and to be ready for it, is all you can hope, to do." Toni's unsuccessful attempts at teaching for understanding did not go unnoticed by the other preservice teachers, who often remarked to the researcher that watching Toni's failed efforts made them reluctant to stray from traditional, structured lessons. It appears that the other preservice teachers were more focused on avoiding confusion or embarrassment in the classroom than they were on teaching for understanding. Although the other preservice teachers frequently employed student-centered activities, their teaching was procedurally-focused the majority of the time. Although Toni believed that she did not have an adequate knowledge base, including a mathematical and pedagogical background to teach conceptually, she believed that, even if she were initially unsuccessful, it was important for her to persist in implementing student-centered, investigative methods of instruction. Toni implied that predictable, traditional instructional methods do not challenge learners to construct understandings. She revealed that these beliefs encouraged her to craft lessons which were both conceptually-focused and student-centered. During an interview session, Toni described the genesis of a lesson on graphing linear equations:

> What motivated me to do that was um, I had noticed that as a group, as a whole, myself included, uh, falling into that, [pause] preach and teach kind of standard, traditional, into the very thing that I, and it felt like they were starting to fall back on that because we didn't know what else to do and so, I, I was a little uncomfortable with doing what I did.

Toni indicated that her discomfort was a result of the fact that she feared the lesson would go in a direction for which she was not prepared. She worried that she would become confused and unable to continue with the lesson. However, Toni also recognized that, because her goal was to facilitate learner understanding, she decided to plunge ahead with the strategy. Toni was aware of when and how the lesson did become confusing for the learners:

> As I recall, there was, at one point, that things did get a little confused, because I was hearing the wrong thing, one student was saying one thing and another was saying another, and I thought they were saying the same thing, uh, and then when I realized, I think either Mollie or Jamie, pointed out to me that they were using different, different equations, and I had, I had missed that, I thought we were both, everybody was trying to do the same, thing, and I had missed I think the plus [referring to some students using 2x and others using 2x +7] and then one of them pointed that out to me and I went oh, you know, well, ok, I can see the confusion on that, and then I'm like ok, what do I do, then I called everybody back to the table and was kind of, started over, and then went again, and uh, my objective with, with, not, giving them answers.

Toni admitted that, as the lesson proceeded, it was extremely difficult for her not to tell the learners the answers. This became particularly stressful for Toni when the learners became frustrated and got angry at her for not giving them the answers. Toni indicated that she pursued this strategy, in part, because she tended to learn more when she was frustrated and struggling:

> I really do enjoy that, "oh, gosh, what's the deal here, what's going on?", struggle kind of thing, and so, I felt that if I just kept throwing it back in their lap, they might, catch that, um [pause] you know, teachers who come over and just do it for you, I have lab assistant that's doing that right now, uh, I ask her a question and she doesn't, it's like she doesn't even hear my question, she just shows me what to do, and, and I still don't have the answer to my question, you know, and I don't know what she did, cause she did it so fast.

Toni's critique of her lesson on graphing suggests that utilizing student-centered, open-ended, investigative teaching methods is more likely to reveal a teacher's weak understandings than adhering to teacher-centered, traditional methods. Unlike the other preservice teachers who were primarily focused on their about their grades in their education courses, Toni was more concerned that her minimal content knowledge and weak pedagogical knowledge would be exposed in front of the learners and her fellow teachers:

> I wanted to throw things back and, and see where, where they would go, and the part that I was nervous about, was that they would go somewhere that I wasn't prepared to help them out, that's the part that, that, that was like a little scary for me, and it's, I think it's because I don't have as strong a background, in mathematics, in all the content areas, as a lot of people do.

In a subsequent conversation with the researcher about her attempts to teach rational number operations for understanding by utilizing a rectangle model, Toni explained how things became confused:

> Well, what went wrong, I know what went wrong that day is I, I took someone else's idea that they offered and I thought, ok, I'll try that, it sounds like a good idea but it was an alien idea to the way I do things so I wasn't comfortable with it and it's a lesson that I've learned more than once and this time I'm going to try to remember it, stick to what I know I can do, I was trying to do that conceptual thing and I wasn't confident enough with it myself in that form to where I was having a hard time getting it to them and I knew that about three steps of the way in there and like I said it'll be the last time I borrow an idea from someone, even if I think it's a good idea and I don't know how to you know work with it so anyway.

Toni's comment suggests that merely providing a novice teacher with an idea for teaching conceptually is not an adequate strategy. As Toni explained, the concepts that underpinned the suggested instructional strategy were "alien" to her. The strategy failed, in part, because of Toni's weak understandings of the underlying concept, but also because of her lack of confidence in her ability to explain the concept to the learners. Toni recognized that teaching conceptually required more in-depth preparation than she had previously realized. The excerpt that follows is representative of Toni's attempts to teach conceptually by engaging the learners in discussion in order to elicit their conceptual understandings. During the course of the lesson, which was intended to be a review of previously covered topics, Toni attempted to encourage the learners to articulate their understandings. She explained the genesis of the instructional strategy utilized in this lesson:

> Oh, concepts, and I was getting really concerned that they were all saying they understood and they all knew what was going on, and, and the only way that I could think of to actually, assess whether they, really did know, was to, to ask them to tell me, without me helping me out at all, I wanted them to tell me, and then I thought, during the course of the discussion, uh, um, more things would come out and then they would, they would be digging it up, instead of me just saying here, this is what you need to know, and, uh, I felt that was the most, more useful way to do it.

Johnny: I think functions are the process of finding an answer to an equation.[Johnny's comment is followed by laughter from Mollie]
Toni: What does anybody else think about that?
Mollie: [to Johnny, in a sarcastic voice] What'd you say?
Johnny: [to Mollie] What's so funny about that man?
Mollie: [to Johnny, in a mocking voice] The process of finding an answer to an equation?
Johnny: Functions are the process of finding the answer to an equation.
Mollie: [to Johnny] So what? Functions are something you do to get the answer to an equation?
Johnny: Basically.
Mollie: [to Toni] Then what's an equation?
Johnny: It's like
Toni: [to Rianne] Wait a minute, that's your job.
Rianne: No, that's her job.
Toni: No, that's your job.
Johnny: I mean you could also□
Mollie: [to Toni] Well give me an example of an equation.
Jamie: That pizza one.
Mollie: [looking through notes] So, is, is
Toni: How does what you wrote about functions fit in with that equation? How can you? [pause]

Mollie: Is? Toni poses a question which focuses on the personal sense making of the learners:
Toni: Does it make sense using that equation?
Mollie: This? [long pause] [At this point, Toni is clearly struggling to respond to the learners. Once again, the pizza context emerges.]
Toni: Well think about, think about our pizza thing that we did, what was the function there?
Jamie: The amount of pizza. [Toni follows Jamie's response with two more questions which focus on the personal sense making of the learners:]
Toni: Ok, who can come up with an example? [The question that modeled the pizza problem was reintroduced]
Toni: So, uh, uh, as, far as a function, what, what, is the purpose, what does a function, help you do?
Johnny: Get the answer.
Toni: Ok, ok.
Johnny: It's a process in which numbers are being increased, decreased, multiplied, and divided by. [Toni does not utilize Johnny's response to elevate a procedural answer into more conceptual realms]
Toni: That sounds like a good answer, that's what's going on in a function.
Johnny: Yeah.
Toni: Numbers are being, adjusted.
Johnny: That's how you, that's basically how you find the answer, the equation, is just [pause] the numbers [pause] it's the process of being added or subtracted but?
Mollie: That's a function?
Toni: That's a function.
Johnny: Is the actual function?
Toni: That's the machine part of the function, yeah, yeah, it changes what you, what you start with.
Johnny: You want us to know, to know how the problem works, that's how it works, that's the function.
Toni: Yes [pause] so like if you wanted to.
Johnny: Besides just knowing the equation you're doing it, that's what you want us to know.

Toni reflected on the lesson:

> It was a discussion, and that's what I wanted it to be, I didn't want it to be a long drawn out, you know for half the class or anything but, but I really wanted them to ask each other the questions, and I was hoping to just put some things in and then serve as moderator when it started getting off the, off the mark, and I think something I may have done differently would've been to, and I think I had thought about it, but I don't think I actually did it, hand out, a sheet, with each of the words on there, function, and geometry and whatnot, for them to, write down what everybody said and I didn't do that, but I think I will and that's one of the things, they're not the best note takers.

Toni stated that the pre-student teaching experience increased her confidence, "Being able to, make mistakes, and, and, recover, and, scramble, to fix things, that was definitely, a helpful thing, to me, that was one of the most helpful things to me, I think, being able to do that." At the end of the semester, Toni articulated a view of mathematics which reflected a belief in the importance of teaching mathematics to assist learners in developing critical thinking skills. This differed from her earlier view that the most important reason for teaching mathematics was to mentor and motivate learners:

> Oh golly, I think, teaching math is very important because you learn more than just the math, the mathematics when you, when you learn math [pause] ahh, it seems to me they should integrate everything else in with math, you know, because that's you, you do, need to look at it that way, because what, what learning math does for me, and I would hope that it would do it for other people is, is it causes you to think things in ways that, you know, makes you better able to discern what's going on, you know, and you, you can articulate things, better, you can decide what's true and false better, and not just about, mathematical concepts, but, once you learn how this, reasoning, that goes along with math, operates, then you can apply that to, finding out who's, you know, pulling the wool over your eyes, or, stuff like that, there's all kinds of, I think, social things that go along with learning math.

Toni revealed that one of the most important things she learned about teaching from this experience was never to ask learners if they understood. She elaborated on how she would go about determining learners' understandings:

> Never ask them, "Do you understand?", ask them instead [pause] "Tell me what you know" [laughs] and have them, explain, back to you cause if, you, what I've found if we approached it as "Do you understand this?", they'll go, "mmhmm, yeah, yep, get it, mmhmm", but if you ask them, "Tell me what you know, explain to me, how you got your answer" then they, actually took the time to think about it and then they started understanding more of what they were doing, and so that's by far a better way to go.

Melissa

Throughout the course of the study, Melissa expressed confidence in both her mathematical and teaching abilities. Observations of her teaching revealed that she was, in fact, quite comfortable with both the mathematical content areas and the classroom teaching. Melissa's high self-concept in mathematics appears to have been shaped by both her fondness for the subject and her positive experiences with school mathematics:

> I guess I really like it because it's challenging, it's a challenge, it's such a great feeling, I mean I can, I can sit there and work on a math problem all night long, and when I get done, when I get done with it, man, it's just the greatest feeling, if I come to the right answer, or come to some kind of conclusion.

Melissa's passion for mathematics enabled her to avoid becoming discouraged despite experiencing mediocre teaching when she herself was a learner:

> I would say that I've always pretty much had an interest in math, I've always been fascinated with numbers, I guess, I just...I took five years, I ended up taking double, doubling up, so I ended up taking algebra, geometry, algebra 2, and then I took precalculus and math analysis so I would say most of my classes, I really wouldn't say I had any good math teachers when I was in high school, but I didn't realize that until I went to college and then I was just like wow, you know, I mean I, when I went away to college then I had some really good math teachers and I just thought, you know, most of them were, you know basically lecture type classes, yeah just your traditional and it didn't turn me off, no, basically I guess because I've always enjoyed math it didn't affect me in any way.

While Melissa appeared to be very comfortable when teaching, she nevertheless solicited the researcher for suggestions, feedback, and critiques of her lessons. She was also curious as to how the researcher had taught a certain concept, how long it took the researcher to prepare a lesson, and how the researcher dealt with issues of classroom management. During an initial interview, when asked to describe what she believed was the most important reason for teaching mathematics, Melissa responded:

> I think that you need math to do just about anything, I mean, you use math for so many things that people don't even, you don't even realize it and a lot of it is you just don't, (pause) I mean just even being taken advantage of, you know what I mean, you don't, you just wouldn't even know if someone is taking advantage of you if you don't have any basic math skills, you know.

Melissa articulated a utilitarian view that the most important reason for teaching mathematics was to assist learners in acquiring proficiency in the mathematical skills that they would need to survive in life. It appears that Melissa's beliefs about mathematics were shaped, in part, by her own background, "My mom has an elementary school background and most of her jobs have been janitorial and just kind of labor work. My two sisters, neither one of them, they're both housewives." During the course of the study, Melissa explained to the researcher that her mother was a Korean whom her father, an American, had met and married while stationed in Korea with the U.S. military. She shared several stories about how her mother had recently been "ripped off" and "cheated" because of her lack of financial acumen. Melissa's background and experiences appear to have shaped her beliefs about the importance of mathematics for "surviving in life." These beliefs, in turn, influenced her instructional practices. Melissa described the mathematics she taught as "just the basics, you know, enough to really just get them by." Consistent with her belief that the primary reason for teaching mathematics was to provide learners with mathematical skills, Melissa's lessons adhered to a fairly traditional format and her teaching style varied little during the course of the semester. For example, Melissa was careful to provide the learners with step-by-step demonstrations, instructions, and explanations. Melissa's instructional practices also appear to have been influenced by her beliefs about learners. Melissa recognized that low self-confidence played a role in mathematics underachievement, "I think a lot of it just has to do with confidence, I think a lot of people have really low self-confidence and so they kind of put up their guard or they don't leave their, they don't leave an open mind." It appears that Melissa's belief that self-confidence was related to mathematics achievement influenced her instructional practices. Melissa's instructional practices may have been motivated by a desire to enhance the learners' self-confidence by providing them with closed questions to which they could all successfully respond. For example, despite her rather traditional approach to mathematical instruction, Melissa consistently made an effort to provide each learner with individual attention. She frequently circulated among the learners, checking for understanding, and offering one-on-one assistance. Melissa also designed her exploratory lessons to ensure that all learners would be successful. During the graphing, orienteering, and Sketchpad lessons, Melissa gave each learner a task that he/she could successfully complete by following her step-by-step directions. Melissa frequently brought in cookies or candies and "rewarded" the learners with these at the conclusion of her lessons. For example, when the learners returned from the orienteering activity, Melissa had candy bars waiting on each learner's notebook. While Melissa revealed that teaching at the alternative high school was "not as different as I thought it would be," she did note that the learners "require more attention, maybe more explanations." Melissa was extremely patient and willing to offer individual assistance and explanations to the learners. She also appeared to be aware of the learners' discomfort at open-ended questions and structured her lessons to minimize embarrassing the learners or making them feel self-conscious when they were unable to answer questions. For example, rather than probing or pursing learners for a response to her questions, Melissa would answer the questions herself:

> Melissa: Ok, now for the next question, um, what you guys get for the area of this figure, now that shouldn't have been too hard because you actually have the measurements around the side so, what did you multiply together?
> Johnny: I multiplied the base times height. [Rather than encouraging Johnny to continue his explanation, Melissa demonstrates the solution for the learners]
> Melissa: Ok, so, this is G, H, I, J, right? So this is 2.75, this is a rectangle so this is 2.75, and you got 1.43.

Melissa's beliefs about mathematics teachers also appear to have influenced her instructional practices. According to Melissa, when teachers enjoyed teaching mathematics, it created a positive learning environment for students. She believed that her enjoyment of mathematics would "make the students want to do better and make them enjoy the class as well." Melissa also believed that the role of the mathematics teacher was, "To provide positive learning experiences, because so many people have negative attitudes towards math and I know a lot of it has to do with the teachers they've had." Melissa's belief in lecture and traditional methods of mathematics instruction are revealed in the following passage which, in addition to revealing her belief about mathematics instruction, she also expresses doubt about the efficacy of alternative methods and explanations which, she implied, do not always "work":

> I'd say I know my subject pretty well, I feel like I know it pretty well, and I would say that maybe another strength is that I, if they don't get some way, I try and think of another way to, you know you keep saying the same thing over and over again, and trying to get it to them that way, for most of the time, most of the time, I mean that sometimes I don't have another way you know, but I would be open to you know if anyone else had a different way that they wanted to look, this, that, sometimes, you have to be careful with that, then, because sometimes it only works for certain cases, which is why, you know, most people don't use it.

When describing what was easy or difficult for learners about the concept of slope Melissa referred to her lesson where she used masking tape to create a "life-size" Cartesian graph on the classroom floor and the learners used their bodies to mark the coordinates. Melissa's reflection on this lesson suggests a reason why she was reluctant to teach in this manner. Specifically, Melissa believed that learners would have difficulty seeing the mathematical content in the lessons:

> I mean that was a real good way for them to actually visualize, you know, to be able to, to kind of walk around and do that kind of thing, I think that made it easier for them, but um, sometimes I think it's just hard for them because they're, they have a hard time seeing the math in it.

In her role as teacher, Melissa did the majority of talking while the learners provided only brief responses. Her lessons reproduced a traditional dialogue pattern of teacher question, student response, teacher evaluate (Cazden, 1986). However, during a self-critique of her teaching, Melissa made a comment that indicated a desire to encourage student discussion, "As far as discussion, to get them all talking, I was probably just trying to throw the question back at them and make it so it wasn't just a one word answer, I wanted them to kind of do more explaining I think I was trying to focus on the concepts." It appears that Melissa had developed her teaching style prior to embarking upon the pre-student teaching experience. However, perhaps as a result of participating in this experience, Melissa seemed willing to try alternative instructional strategies. Melissa experimented with open-ended activities during the semester-long mathematics course including having the learners: (1) use the life-size Cartesian graph to explore the concept of slope; (2) discover pi by measuring bottles, clocks, and peanut butter cups; and, (3) engage in an orienteering exercise to review angle measures. Despite exploring alternative teaching strategies, Melissa revealed that the teaching experience had not changed her instructional practices but, instead, allowed her to hone her "time management skills". Melissa made several comments which suggest she utilized exploratory activities in order to provide the learners with a change of pace and prevent boredom and lethargy by "getting them up and moving." In revealing what she believed was a weakness in her teaching, Melissa alluded to the fact that it was easier for her to provide the learners with answers rather than allowing them time to discover solutions on their own. This is consistent with her uneasiness about the open-ended, exploratory methods espoused by the education classes:

> Maybe I don't leave enough time for them to answer my questions, yeah, so, [pause] and another thing too, is maybe, sometimes it makes me a little afraid to try and maybe put them on the spot so instead of like leaving that silence there I'll go ahead and answer or uh, or there's probably more I could do where I could get them more directly involved with it, and maybe I don't just do it because it's almost easier just to do it myself.

While what follows is but one excerpt from eight transcripts of lessons taught by Melissa, it is representative of her teaching style. For example, as Melissa recognized in her previous comment, she did the majority of talking and explaining:

> Melissa: Ok, if I draw an angle [draws an angle] I want to find the measure of that angle right there, who can tell me what that is?
> Mollie: I can.
> Melissa: Ok.
> Mollie: That is about 150
> Melissa: 150?
> Mollie: Mmhmm [Rather than allowing one of the learners to describe how they arrived at an angle measure, Melissa demonstrated a procedure herself]
> Melissa: Ok, everybody see how she gets that? If I draw one like [draws another angle] this, ok, this is a 90 degree angle, like this, so if you have anything that's less than this, it's probably going to be? [At this point Melissa noticed Jamie was having a problem with the protractor. She briefly turned her attention to assisting Jamie]
> Melissa: You're going to be using the bottom scale, you don't have one? Oh, ok, I would probably turn it upside down, or put it backgrounds, and then it measures that way, ok [Melissa then turned her attention back to the whole class]\
> Melissa: I want you to take out a piece of paper, out of your notebooks, and draw these angles, an angle of 145 degrees, and draw one that's 25 degrees. [Melissa writes these instructions on the whiteboard]

While the learners were working on drawing angles, Melissa circulated to offer individualized help and check student work. Once again, rather than asking the learners to explain their strategies, Melissa showed the learners:

> Melissa: Ok, I'm just going to do the 145 degree one, so if I start here and then, ok, this is, this is my point where I'm going to put my protractor, so if I do that [demonstrates on board] then 145, it's going to be more than 90, I'm going to read it this way, I'm going to go to 145, which is about right here, and then I would just mark, there's the 145 degree angle, and then the 25, we're going to do the 25, you just uh, put it here, it's less than 90, the 90 degree angle, and that's going to be right here, and you just go like that, ok? So remember we went over the little

activity where we had the triangles and you ripped off the corners? So what did we say about when you add up the angles of the triangles what do they add up to be? [Jamie and Johnny both answer " 90 degrees"]
Melissa: What angles? What is it?
Johnny: 90?
Melissa: 90 or 180?
Johnny: 180.
Melissa: 180, ok, so, I have a triangle like this [draws on board] and I have, this is 40 degrees, this is, 85 degrees, how am I going to tell that angle, go ahead and do that one on your paper, and you don't have to actually measure them, because if you know the measure of 2 angles, then you can find the measure of the third angle, so how would you do that?
Mollie: Add them together and subtract them, the total from 180?
Melissa: Right. [As the learners worked on the problem, Melissa circulated to check the students' work and offer individual assistance. She once again provided the learners' with an explanation rather than allowing them to articulate their own strategies]
Melissa: Ok, remember we said that they add up to 180, so that if you want to find this angle measure, what we're going to is, we're going to add these 2 up and we get 40 plus 85, 125, and then subtract that from 180, so then 180 minus 125, is 55. You guys all feel comfortable with that? Ok.

Melissa expressed tremendous admiration for Mr. Blake, an eighth-grade mathematics teacher she observed during a six-week practicum earlier in her teacher preparation program. She noted that, while he had developed a positive relationship with his learners and was well-liked by the students, he utilized rather traditional instructional practices. Melissa was comfortable with her own style, which appeared to be validated by her recollections of Mr. Blake. The ideas and strategies that she observed in the classroom, did little to change Melissa's instructional practices. She noted:

> You have to go with your own teaching style too, I mean, you could learn all these things in these classes, but they aren't all necessarily going to work for you either, I mean, you have to go by what works for you, if it doesn't work for you, it's going to show real quick.

At the conclusion of the study, Melissa amended her description to include:

> Someone whose just going to be, open, but I think that a good math teacher is also going to, you know, let students realize that there's more than one way to solve a problem, they're going to be, they're going to, I don't know, I think they would try and think of ways that, to present math that's a little more interesting than traditional methods.

At the conclusion of the semester, Melissa reflected on what she learned from participating in the pre-student teaching experience:

> I think that when we first get out there, we're going to be making, I'm sure lots of mistakes and we're, but we're going to learn from those mistakes and

> one of the things that I think is really important is when you have a lesson you have to, you aren't going to be able to anticipate every direction it's going to take but you have to try and figure out you know, where it might be able to go because if you're just going to follow that lesson plan and you're going to get up there and start teaching it, and you know, someone starts taking it over here, you're going to be looking at this lesson plan trying to figure out where you are and you're going to get lost and then you know, it's just, it can end up just confusing them, and what could have been a really good lesson just kind of flops on you, so, I just think that that's really something that's going to come with experience and you're going to make mistakes.

Observations of the mathematics class indicate that, although she dabbled with open-ended, exploratory activities, Melissa's teaching varied little in style from the beginning to the end of the semester. Melissa indicated that participating in the teaching experience did not change her beliefs about teaching mathematics. However she did reveal that the experience enhanced her teaching, "Just as far as like time management, maybe better planning, uh, that kind of, you know, more flexibility."

Melissa revealed that she felt a pressure from her education classes to implement more "hands-on, fun activities" into every mathematics lesson:

> I think that's one of the things that you have to remember, is, I mean that even though it's nice to have activities, there's just no way you can do that the whole time, I mean you have to have a, you have to have a variety, I think, of stuff, and that's why it really bothers me when people, think that you have to have an activity for everything, cause some things are just better learned, in a traditional method, and other things are better learned the other way.

Mark

At the beginning of the study, Mark had a high self-concept in both mathematics and teaching. He described the teaching he did with the severe behavior disorder learners as "a great experience." Conversations with Mark indicate that his high mathematics self-concept was shaped by his positive experiences with high school mathematics as well as by his love of the subject:

> It came easily to me, it was ok, pretty positive. I was pretty proud of the fact that I was good in math and I wanted to skip years ahead, I went a year ahead when I was in grade school and when I came, when I entered my freshman year, geometry was closed out for some reason so I took intermediate algebra and I already had that so the next year, my sophomore year, I took geometry and precalc, out of my own choice just to get ahead faster. In college, either I liked the prof and I enjoyed the material and I studied really hard just because I wanted to or if I didn't I would really skate.

Consistent with explanations of his own early educational experiences, Mark maintained that success in mathematics was a result of a natural talent for mathematics. Mark indicated that this belief had its origins in his early educational experiences, when he attended a school where each individual learner's "talents" were identified and cultivated. As a result, Mark believed that he had a natural ability for mathematics but not for English:

> I think people have natural affinities for different things and there's also a nurture that goes on, nature, nurture, so like I think, I think I'm like naturally better at math than at English, I'm not very, I'm not a very eloquent speaker and I don't, I don't write the best essays and, along with that I went to an alternative school my first few years and the program for that was fostering natural development so I liked math and they really pushed me in math, I didn't like English so I didn't get pushed very hard in English, not pushed but, encouraged, I pushed myself, you know, at work time, it was what do you like, I like math so I did math, and uh, it was a really good environment for me to get better in what I was good at and a poor environment for me to get better in what I'm not so good at.

Mark was the only one of the preservice teachers who consistently articulated a personal desire to go beyond traditional definitions to explore the conceptual underpinnings of mathematics:

> I think I really made an effort in high school to understand math, you know backwards and forwards so something like the unit circle, really not only understanding how to find the sine and memorize the little table like most people did but understand the process that was going and everything involved in that, once you do it like that it really sticks in your head, you know, so, I think, I just, I can pick up that kind of thing, the logic of math really quickly.

However, as the semester progressed Mark commented, "I realize how much I don't know about math." Mark's desire to understand the conceptual underpinnings of mathematical content areas was reflected in his attempts to teach for understanding. He clearly enjoyed contemplating abstract concepts and made frequent attempts to provoke discussions of mathematics and physics phenomena:

> That's the type of stuff that always got me interested in math, it's, it's, I like this concrete stuff, but when you start going into stuff that's not, that you can't nail it down and you have to use more of what you know already, that's what makes me want to learn.

Mark's instructional practices were influenced, in part, by his prior experience with and perceptions of students on the margins:

> Oh, yeah, I definitely think so, some kids, for some of these students, if they get certain types of thinking and certain types of learning that don't fit with the traditional school setting, definitely they become marginalized, and I

> think that some of those students if you have the resources to give them, and I don't know if they do at this school, but, they can learn, they can be the best students, learning more than anyone but they can't do it by sitting and obeying rules, listening, just listening, some people can't listen, they need hands on, they need to experience things.

Mark's teaching self-concept was high at the beginning of the study. He expressed a high level of confidence in his teaching abilities which appeared to be a result of positive experiences teaching marginalized students at his previous position. However, as the semester progressed, Mark began to doubt his teaching abilities. He consulted with the researcher on numerous occasions, soliciting advice and feedback. Mark was the most open of the five teachers in critiquing his teaching and identifying and describing his weaknesses:

> There's a lot I could do better, I could make it a lot more cohesive, more confidence would help, I have some but I could use some more, um, I'm, it's hard for me to know but I feel like, there's more I could be doing, I don't know [pause] but a lot of it is uncertainty and too much, too much processing going on up there, I think that if I did this for a few years, a lot of that processing would go away and it would just be a clear thought process for me and I could keep things a lot more coherent.

Mark elaborated on this thought by explaining that when he initially observed a behavior, such as inattentiveness on the part of a learner, because he hadn't yet developed a strategy for managing it, he would ignore that behavior. He realized that, "Then after you deal with it, figure out a way that it, you know that works once and that's not a problem at all any more, and it frees up your mind, less distraction." As the semester progressed, Mark's low teaching self-concept became more apparent and he revealed that his low self-concept was a factor in his decision to put his plans to teach on hold. He revealed that he was grappling with ways to take the ideas discussed in his education classes and implement them in the classroom. Throughout the semester, Mark continued to be troubled by this struggle:

> Mark: I'm having trouble formulating opinions on what should be done. I feel like I'm floundering, I don't know, how do you know what to do when you don't know, when you don't have enough experience to back up your opinions. I have a feeling what should be done but then again I find myself, like especially setting up the lessons, it looks you know, we're working it out and I think everyone has good ideas, but then again I look at it and I go, well is this right, I don't really know what is right, does that make sense?
> Kevin: What's right is what works.
> Mark: Right, but I don't know what works because I haven't done it.

Although Mark had previously taught mathematics, it was primarily as a teacher's aide rather than as a classroom teacher. Consequently, Mark was accustomed to one-on-one instruction rather than whole class instruction. After his early attempts to teach in front of the classroom, Mark confessed that he was extremely uncomfortable and began to sit with the learners. He revealed that he was more comfortable sitting side-by-side with the learners rather than standing in front of the classroom. Observations confirmed that Mark appeared uncomfortable and awkward in front of the classroom. As he observed a video clip of his teaching, Mark noted how his arms were folded tightly across his chest as he stood in front of the classroom. This discomfort was consistent throughout the semester. Conversely, Mark seemed at ease when he was sitting and working with the learners. These personal traits were also evident in his interactions with the preservice teachers. Mark was the peacemaker who attempted to maintain harmony and minimize conflicts that arose between the preservice teachers over the course of the semester. Mark explained that his personality influenced his teaching style:

> You've just got to be yourself out there, I think, if you're genuine, people will know it, um, I'm uh, kind of a, a weakness for me in all aspects of my life is that I'm a pleaser person, I want people to be happy with me, and so, if like, Wendi didn't want to do certain things, sometimes I'm just going to say, well this is what we're doing and uh, I don't want people to be angry with me or whatever, so I try not to take, I don't like confrontation, and that's going to happen, I'm getting better with that, but, that could be a problem with me, is letting the students control too much [pause] I think I may have control over it, ownership and some healing power, self-respect, but then again, you don't want to be a chump either [laughs] it's a fine line [laughs] and what is true for me might be, a good balance for me might, other people might look at me as being a chump but then that's just the way it is.

Mark displayed a fairly nontraditional teaching style throughout much of the study. He developed several interesting instructional practices that he believed would encourage learners to think about the conceptual underpinnings of the mathematical topics. However, these were not always implemented in the classroom. Mark often adjusted his planned instruction, noting "the students are demanding the formulas." On several occasions he remarked to the researcher that other preservice teachers were emphasizing mastery of formulas in their lessons. For example, Mark cited the geometry lessons as an example of the preservice teachers emphasizing formulas rather than the relationships between and characteristics of polygons. Mark believed this focus encouraged the students to memorize algorithms without meaning, "There's no thought involved, it's just, you know, they're doing arithmetic, not even that, it's just a bunch of numbers." Mark's own limited teaching experience also played a role in preventing him from implementing these strategies as he originally conceived them:

> I was really trying to do that more, more activities that they could get involved in, but I found once I was up there I was a little bit afraid to just kind of let it go, if that makes sense so I was always throwing in something or correcting them, and then I'd think oh I shouldn't have corrected them right

> away. I keep interjecting because I want it to go well but then that kind of spoils the whole purpose of it. I know I've brought in a lot of preconceptions about teaching, about what I'd being going through, and not even close, you know, so, I, I had some ideas of what I knew teaching was about and things, and I really didn't.

At the conclusion of the study, Mark indicated that the most important reason for teaching mathematics was "The world's changing, and, we're not going to need a bunch of little factory workers any more, we need people who can think." Mark revealed that his beliefs about math had not changed as a result of participating in the pre-student teaching experience. Conversations with Mark suggest that his beliefs about mathematics were shaped, in part, by his experiences as a mathematics learner. However, Mark also revealed that he thought a lot about physics and mathematics in his spare time. Of the five preservice teachers who participated in the study, Mark was the only one who mentioned that he reflected on mathematics in this way. He indicated that his time away from the university motivated him to consider mathematics in a way that differed from what he had been traditionally taught. Mark articulated a belief that learning mathematics entailed more than merely mastering a sequence of steps, "If you know how to do something real quick, by a couple of steps, then you really haven't learned it whereas if you know what's going on...then you're learning math." However, after observing the learners' weak computational skills, Mark offered the following comment:

> I'd still like to have some of the, just number crunching problems in there for, one reason, is just to see if they can do it, some of them, I'm not sure they can, and, if they can't do that, then they can't really do anything, no, I mean, no, I mean, well you might have some concepts but you also need to be able to just get the answer, that's important too.

Mark expressed a belief that a variety of factors were responsible for learners underachieving in mathematics, "Motivation is a real key and natural ability has an effect, but I think, I know people that, that don't pick it up as quick as others but, they can learn it just as well, depending on the school." During the final interview, Mark stated that, while some learners possess more aptitude for mathematics, he believed that anyone with enough determination could learn it. He identified two kinds of learners who were likely to underachieve mathematics: (1) those who had struggled with mathematics in the past; and, (2) learners who had fallen behind, because "once you fall behind there's no good system to catch up." Both these categories of learners were represented in the mathematics class at the alternative high school. It is possible that Mark's experiences with these learners may have mediated his beliefs about "natural ability," enabling him to identify other reasons for mathematics underachievement. Mark explained that the role of the mathematics teacher is to motivate learners' interest in mathematics "so that they start thinking about it." Observations of Mark's teaching strategies revealed that he spent a great deal of time sitting with the learners and working through open-ended problems with them. This strategy appears to have resulted, in part, by Mark's belief about teaching mathematics, "As much as possible, be like that, thinking and acting with them, rather than showing it to them but a lot of times you have to show it to them [laughs] you can do that." Mark expressed a belief that an ability to relate to other people and "understand where they're at" was the most important characteristic of a mathematics teacher. However, he also noted that teaching students on the margins was different because of "bringing stuff home with you, maybe you do that in a regular class, I don't know, worrying about them." Mark also recognized the importance of possessing an enthusiasm both for mathematics and for teaching people:

> Discovering math with the students, even if it's something you know, doing it along with them, learning it, not just going in there and teaching, but have it be enjoyable for the teacher when the student finally figures it out and sharing that with the student, has got to be a good characteristic right there.

Observations of Mark in the classroom and in planning sessions revealed that he possessed these characteristics. Classroom observations confirm that Mark genuinely enjoyed exploring mathematics concepts with the learners. He also shared the learners' enthusiasm when they were able to successfully work their way through mathematical tasks. Mark's beliefs that the characteristics of a good mathematics teacher were someone with enthusiasm for mathematics and teaching mathematics remained unchanged at the conclusion of the study. He also held to his belief that a good mathematics teacher "relates to the students well." The following transcript of a lesson on linear functions is representative of Mark's teaching style and highlights the instructional practices he utilized in the classroom. For example, in addition to bringing in manipulatives, throughout the lesson Mark sat with the learners and engaged them in the activity. He also recognized and responded to each learner's comments. Mark began the lesson by sitting at a large table with the learners and bringing out two envelopes filled with candies:

> Mark: All right, so, I've got two envelopes and in each one there's some candy, there's "m & m's" and I have a total of 7, there's 7 in these two [envelopes], I've got some more for later.
> Wendi: You have 4 in one and 3 in the other, or 5 in one, or 2 in the other.
> Mark: Yeah, there's some possibilities, yeah. [The learners continued to name possibilities.]
> Mark: So, so my question is how could you graph some of those?
> Wendi: You have none in one.
> Mark: Some of those answers you just came up with.
> Mollie: Graph them.
> Mark: Find some of the answers that you could get.
> Jamie: Can we feel the envelopes?
> Mark: No feeling the envelopes, no, that's, that's a good thought though. [Students worked on problem and then shared their results with Mark who, in an

effort to encourage collaboration, Mark attempts to have the learners work in small groups]
Mark: Graph some of the answers that you guys got, you guys can work in groups if you want, maybe Wendi and Mollie can work together and you 3 can work together, so you got one answer, what was one answer you came up with?
Mollie: Find a way to keep track of your answers.
Wendi: 5 + 2, 3 + 4, 6 + 1
Mollie: You just do them.
Mark: So, 5 and 2, 3 and 4, 6 and 1, you could graph them on the x, y, so make your x, y plane, that's a good place to start, one's x and one's y. [The learners are very engaged. However, at this point they are merely using arithmetic ideas to find the points and then plot them]
Wendi: This would be x right here, right?
Mark: That's good.
Wendi: So it's that line?
Johnny: How many of them do you want us to do, all the combinations?
Mark: Sure, show me a couple of them and then, and then graph all of them.
Mollie: Should we do 0 and 7?
Jamie: You could have paper in one of these [In an effort to encourage the learners to utilize fractions or decimals, Mark makes a suggestion]
Mark: Sure, what if I cut one in half with an x-acto knife?
Jamie: Oh, um, you're getting tricky on us.
Wendi: Don't even start with that right now.
Jamie: I only have 10 little slots.
[The learners continued graphing their combinations.]
Mollie: 3 ½, should I use that? 3 ½?
Mark: Sure, well what about?
Wendi: Don't even get me started on halves, please
Jamie: There's one little line, it just kept moving.
Mark: I think just do it on paper, Jamie [pause] ok, well do you guys notice about the, the graphs you made? [Because Mark's lesson is focused on understanding, at this point the learners are able to discover the patterns in the points]
Mollie: It's a line
Mark: It's a line, right? So what about the points, in between the points, what do you think happens in between those points that you made?
Mollie: Uh, the numbers get bigger
Mark: Yeah, there's,
Johnny: Halves and quarters
Mark: Right, right
Johnny: Sixteenths
Mark: So if you use your, I think you already did that didn't you, you can draw a line through them and all those will be answers to this, right [Mark then brought out two more envelopes filled with different candies and asked the learners to graph possible solutions to the linear equations $x - y = 2$ and $1/4\ x + 2$. As the learners began to work on those equations, Mark showed the learners the equation they had just worked with]
Mark: So the first one we got was uh, $x + y = 7$. [writes on board]

It appears that Mark's pre-student teaching experiences at the alternative high school affirmed his belief that investigative activities may be of value for "some students." When asked to describe if his beliefs about teaching mathematics had changed as a result of his experiences at Transcend, Mark made a statement which was precisely what Melissa said at the conclusion of the study: "Yeah, I think there, I think there is some value to doing hands-on stuff, for some students, and that you should do both, the hands-on stuff and the more traditional stuff." However, each of these preservice teachers enacted their beliefs in very different ways. Mark's response suggests that he was not entirely convinced of the value of using hands-on, exploratory activities to engage learners and provide alternative ways of visualizing mathematical concepts. This conclusion is supported by comments made by Mark throughout the semester in which he revealed that he often employed teaching strategies which he believed would engage the students in exploring mathematics concepts through real world connections. Mark indicated that he declined to use traditional methods, even when he believed they would be more effective. At the conclusion of the teaching experience Mark explained that, for him, it was a rewarding experience, "I didn't realize I had a lot that I don't know about, about different things, I think you only learn it by trying it out, seeing how it goes." Mark also explained that participating in the teaching experience at the alternative high school assisted him in developing a teaching style and confirmed his assertion that he enjoyed engaging learners in discussion rather than lecturing:

> I think I kind of developed, started to develop a style, I mean I didn't have one before, but I do like working in discuss, I like discussing, sitting around in a circle like that, because, that's what I did for most of the last half, and it's just comfortable for me, rather than lecturing, it's just not my, personality, to be lecturing. I started realizing, well, hey, I'm a novice and I'm going to make mistakes, and I got to do the best I can, and let the mistakes be, it's ok, make them ok with me, you know?"

Effect of the Teaching Experience on the Preservice Teachers

Engaging in the pre-student teaching experience impacted the preservice teachers in several ways. Over the course of the semester, there were changes to their self-concepts, perceptions of at-risk learners, and instructional practices. Although they impacted instruction in various ways, for the most part, the preservice teachers' beliefs remained unchanged by the teaching experience. It appears that their beliefs were rooted in their prior educational experiences. However, as a result of participating in the teaching experience, several of the preservice teachers began to rethink several of their ideas about teaching and learning mathematics. In addition to realizing the importance of multiple methods and mathematical discussions in instruction, they began to recognize the importance of understanding the conceptual underpinnings of the mathematics being taught. The teaching experience also appeared to have had a positive effect upon the preservice teachers' self-concepts and their perceptions of marginalized learners. These, in turn, impacted instructional practices. In an effort to determine the effect of the teaching experience on instructional practice, this chapter examined several issues related to instruction. Those issues which appeared to have had an impact on instructional practice include beliefs, backgrounds and experiences with school mathematics, perceptions of learners, and self-concepts. The learning patterns of the students also appear to have impacted instructional practice. For example, the learners' reluctance to engage in mathematical discussions and their resistance to open-ended exploratory activities often encouraged the preservice teachers to abandon conceptually-focused instruction. There were instances in which the preservice teachers complied with the learners' demands for standard algorithms. During a discussion which occurred between the preservice teachers and learners which immediately followed a conceptually-focused lesson, Johnny explained, "I don't like the whole concept thing. I never been learned the concepts before." A discussion of the impact of the teaching experience on the learners is provided in the following chapter.

For Exploration and Reflection

1. How do the mathematics lessons in this chapter differ from the mathematics lessons you experienced as a student? How are they the same?
2. What qualities do you observe in each of the preservice teachers? Provide examples.
3. What strategies could Brad have employed to assist the students in constructing mathematical concepts? What are some questions that he might have asked?
4. What manipulatives would you employ to enhance student understanding within these lessons? What kinds of technology-based activities could you integrate into the lessons to support the learning and understanding of mathematics concepts?
5. When Mark questioned the students about circles and pi, what was he trying to accomplish? What else might have he done?
6. What are the advantages and disadvantages of the strategies employed by Toni and Mark? What were they trying to accomplish?
7. During lesson in which Mark and Toni attempted to teach the multiplication of positive and negative numbers with understanding, the students were reluctant to engage in a discussion of their strategies. What are some ways that you would have attempted to engage them in explanations of their strategies?
8. Do you think that the instructional strategies employed by Melissa are helpful or hurtful to students' developing understanding of mathematics? In what ways?
9. Discuss Kevin's use of snowboarding in his lesson on angle measurement. If your students are unfamiliar with snowboarding, what other examples could you use to connect the mathematical concepts to their own cultural and community experiences?

10. What are the advantages and disadvantages of the way in which Toni responded to Jamie and Johnny during her lesson on area? Would you have responded differently? Explain.
11. Comment on the appropriateness of Toni's function lesson for this group of students. What would you do differently? Support your response.
12. What types of questions does Melissa ask? Do these sorts of questions further mathematical understanding and learning? Provide examples.
13. What other activities related to the mathematics content in this chapter could you plan? What mathematical connections could be emphasized? What mathematics might emerge?
14. What were the characteristics of each of the preservice teachers' interactions with the students? What types of questions did they ask? What statements did they make?

Chapter Five

The Learners

The learners profiled in this chapter were introduced in Chapter Four. All five learners elected to enroll in the mathematics course taught by preservice teachers at the alternative high school. Although the learners are demographically diverse, all were identified as marginalized and disconnected and had been labeled "at-risk." This chapter will present information on these learners' beliefs, self-concept, learning patterns, and mathematical development.

Beliefs

Research supports the assertion that learners' beliefs are the "invisible hand" operating in mathematics instruction. For example, learners who believe that learning mathematics is a matter of mastering a battery of procedures may resist instructional practice which encourage reasoning about or understanding mathematical ideas (Schoenfeld, 1985; Borasi, 1990). Therefore, because learners' beliefs play a role in mathematics teaching and learning, the purpose of this section is to examine the beliefs held by the group of learners who participated in this study. The learners' beliefs about mathematics and mathematics learning were categorized and clustered into the following groups: (1) mathematics as a tool consisting of a body of procedures vs. mathematics as a system of conceptual knowledge; (2) learning by listening vs. learning by discovery; (3) focus on the individual person vs. focus on the content; (4) one solution method vs. multiple solution methods.

Mathematics as a Tool Consisting of a Body of Procedures vs. Mathematics as a System of Conceptual Knowledge

Nearly all of learners shared a belief with the preservice teachers that mathematics allowed for creativity, exploration, and argument. Mollie was the only participant whose responses indicated she was uncertain about her beliefs in this area. The learners' belief that mathematics allowed for creativity, exploration, and argument suggests that they did not regard mathematics as a static body of knowledge to be memorized and applied. This belief contrasts with interview and classroom data that suggest that the learners believed success in mathematics was a matter of memorizing and executing a battery of procedures demonstrated by the teacher.

Learning by Listening vs. Learning by Discovery

Throughout the study, the learners held to their belief that mathematics was rule-based. As such, they believed that success in mathematics was a matter of learning and applying a series of formulas and procedures. Johnny's request to the preservice teachers is representative of the learners' beliefs, "I don't care about the concepts, just give me the formula, man." As noted in the previous sections on the preservice teachers, their inexperience with teaching for understanding and conceptually-based teaching resulted in an emphasis on procedurally-based mathematics learning. It is possible that, because the preservice teachers' instruction was dominated by procedural methods, it served to reinforce the learners' prior beliefs that learning mathematics was a process of mastering a series of rules, formulas, and procedures.

Focus on the Individual Person vs. Focus on the Content

For the most part, the learners commented on the theme of mathematics learning in a non-mathematical framework. Rather than focusing on the learning itself, they spoke of issues of caring and seeing things from the perspective of the learners. These beliefs appear to have been shaped by their own negative experiences with school mathematics and the teachers who taught those courses. All of the learners shared accounts of how their personal problems affected their mathematics learning. Consequently, they identified "good" teachers as those who were able to facilitate their learning in spite of personal difficulties. As Jamie explained:

> I took math in eighth-grade and I had a great teacher, he just really cared about the students. He took a lot of time with me, I was having like a lot of problems and he really cared, he took the time, he'd take me out of class, and he just knew when things were going bad for me he was just a really good teacher.

In an earlier discussion about this same teacher, Jamie explained, "He knew a lot of stuff about how I liked to learn things, that's a very important thing, is, for you to be aware of how you learn." She also noted that, "He just made every problem that you did seem like this is something that you can relate to."

One Solution Method vs. Multiple Solution Methods

With the exception of Johnny, the learners all agreed that they learned mathematics best by employing a variety of strategies. However, they also believed that a good mathematics teacher would show learners exactly how to answer problems on a mathematics test, which was in direct contrast to the preservice teachers. Johnny noted that the multiple methods were confusing:

> They kind of conflict with each other it seems, you know, one person teaching one way and the other one teaching the other, and, when you do it one way, you're not going to try and do it the other because you already know the one way.

Self-Concept

The learners began the semester with low mathematics self-concepts. Interview and classroom observations suggest that these low self-concepts were a result of their prior negative experiences with school mathematics. It is also possible that the learners' low mathematical self-concepts were related to their belief that learning mathematics was a matter of memorizing a collection of mathematical formulas. Their inability to recall formulas and rules may have reinforced their low mathematics self-concepts. Conversely, the encouragement and individual attention the learners received from the preservice teachers may have assisted them in constructing high mathematical self-concepts. Participating in the teaching and learning experience with the preservice teachers appears to have had a positive effect upon the learners' self-concepts. The instructional practices designed and implemented by the preservice teachers had positive effects on the learners' mathematical development. However, it appears that the preservice teachers were also extremely effective in enhancing the self-concepts of this particular group of learners. In particular, because of success on narrowly defined procedural tasks and the individual attention received while working them, the learners seemed to gain a greater sense of confidence in regard to mathematics. For example, Melissa's and Brad's fairly traditional teaching styles enabled the learners to rotely follow along with minimal cognitive demands. When the learners were successful at solving these procedural tasks, they felt as if they were achieving in mathematics and their self-concepts were enhanced. The mathematical learning environment at the alternative high school appears to have facilitated social interaction between teacher and learner. Learners were encouraged to speak without fear of humiliation or reprisal. Moreover, the personal bond that developed between the two groups appears to have enhanced the learners' perceptions of themselves as mathematics learners and, in turn, improved their self-concepts. These results were reinforced by data from questionnaire items designed to explore the learners' mathematics self-concept by examining their attributions of success and failure. Nearly all of the preservice teachers and learners attributed success in mathematics to hard work. However, no patterns were found with regard to learner comments on attributions of success and failure in mathematics. Consistent with the learners' rather low mathematics self-concept, with the exception of Jamie, all of the learners suggested that when they got a different answer to a problem than the other learners, it was possible that they had made a mistake. Jamie indicated that when she got an answer to a problem that was different from the answers of others, they could both be right. This suggests that Jamie did not view mathematics as rigid or composed of universal truths.

Mathematical Development

While the learners did make some gains in the area of arithmetic skills, their progress does not appear to be consistent. Although there were instances in which the learners appeared to have improved their arithmetic skills, there were other cases in which the learners got the formulas confused and, as a result, applied incorrect formulas. While on some occasions the learners were able to accurately apply the appropriate formulas to solve equations, at other times they had difficulty recalling the formulas themselves. This observation is consistent with the views of Hiebert and Lefevre (1986) who assert that the forgetting of mathematical procedures is related to a lack of conceptual mathematical knowledge. In this particular study, it is possible that the learners' lack of conceptual knowledge played a role in their inability to recall mathematical algorithms, rules, and procedures. Observations of the mathematics classes suggest that the preservice teachers did not provide the learners with sufficient practice to master mathematical procedures. More importantly, the procedural focus of the lessons appears to have prohibited the learners from constructing their own understandings of the mathematical procedures. For example, the preservice teachers taught the learners the Pythagorean Theorem and the formula for finding the area of a triangle utilizing step-by-step instructional approaches in which the mathematical truth was ratified by the preservice teachers (Lampert, 1990). The learners were not provided with opportunities to construct their own understandings of these mathematical procedures. Instead, the learners followed a sequence of steps laid out by the teacher:

> Kevin: Now there's a neat little property with the, with the uh right triangle, you could find out different sides, when you have a right triangle, let's say we have lengths of, 3, this one will be 5, anyone want to guess about what that length might be? [draws a right triangle on the board]
> Jamie: 8.
> Kevin: 8? You'd probably be pretty close.
> Wendi: It'd be a good [pause]
> Mollie: It's 5.
> Wendi: Well, because you add 3 and 5.

Kevin: Remember how angles are proportionate to the sides? So, this side's, this angle's going to be
Wendi: Kevin, he's trying to tell you something. [referring to Brad, who tells Kevin to change the 5 to a 4]
Kevin: Ok, so that's 4, so what would this be, 7 or so, do you think? Ok, now the way you can find this, is a neat little Theorem that our friends the Pythagoreans came up with, known as the Pythagorean Theorem, you guys ever heard of the Pythagorean Theorem?
Wendi: No.
Kevin: Remember how to do those squares, Jamie? You have 3 squared.
Jamie: 9.
Kevin: Right, so for this triangle, you know that a squared is going to be 9, and b squared's going to be?
Wendi: 16.
Kevin: 16, and that's going to equal, what's that?
Wendi: 25.
Kevin: 25, equals c squared, ok.
Mollie: Oh, so that doesn't that?
Kevin: What do you do to get c by itself? This is?
Mollie: You divide? [Kevin led the learners through several more calculations involving squares and square roots and then summarized the importance of the Pythagorean Theorem]
Kevin: Right, right, so the square root of 25 is, you take the square root of it [Rianne and Mollie both answer 5] right, 5, so we know that this side right here is equal to 5, why don't we try another one of those? This is, once you understand how to square things, and take the square root of things, it's a pretty, it's a pretty handy little Theorem to have, because you can find the different lengths, find different lengths of right triangles, why don't we try one more?

Although triangles, including right triangles, were incorporated into six subsequent lessons, the Pythagorean Theorem was reviewed in only one of those lessons. Once again the preservice teacher utilized a step-by-step approach that emphasized the formula and did not provide the learners with an opportunity to reflect upon their own understandings of the Pythagorean Theorem:

Melissa: Does any one remember what we use to find c? [refers to a right triangle with a hypotenuse labeled "c"]
Johnny: A squared plus b squared equals c squared? Is that what you?
Melissa: Right, a squared plus b squared equals c squared [writes the formula on the board] remember what the name of that's called?
Johnny: Uhh, I remember the formula, [pause] Pythagorean Theorem.
Melissa: Good! Pythagorean Theorem, remember that we said that, in this case, the x is equal to, if c is equal to x, that means that this is going to be either a or b, whichever, however you decide, and this is going to be the opposite, it's going to be c, so if we let 7 equal a, let a equal to 7, and b is equal to 16, and who can tell me how to set up the formula? What am I going to write?
Jamie: 49.
Melissa: 7 squared.
Jamie: And 16 squared.

Melissa: Plus 16 squared [writing on board] is x squared, good, all right. [Melissa guides the learners through the final steps]

During a mathematics test, none of the learners were able to utilize the Pythagorean Theorem to find the length of the diagonal of a rectangle with sides of twelve and five units. Instead, four of the five learners multiplied the two lengths and divided the result by two. The fifth learner, Mollie, did not have a mathematical response to the problem. She wrote on her paper, "At first I thought I knew a way to solve it, but I thought about it and that's a whole different subject." The learners who attempted to solve the problem were unable to recognize that the procedure they utilized did not give them the length of the diagonal but, instead, was the formula for the area of the right triangle, or one-half the area of the rectangle. They did not perceive that their answer, thirty, was an unreasonable value for the diagonal of a rectangle with sides of twelve and five. The data suggest that the procedural focus of the lessons did not provide the learners with an opportunity to interpret the procedure in relation to their own prior knowledge and to reconstruct it in their mind (Hatano, 1996). It appears that the learners' misunderstandings of the underlying concepts interfered with their ability to solve the problem. In a reflection on one of his lessons, Mark's comments suggest that he recognized that an emphasis on the memorization of formulas, without a relational understanding of the underlying concepts, prevented learners from solving problems when the formulas were forgotten. It appears that, when the learners were unable to recall the correct formula, their weak conceptual understandings prevented them from extending their procedural knowledge to solve novel problems. He stated:

> This whole thing [the lesson] is trying to understand the area of the circle, where the area of the circle formula comes from, which they already did earlier, [pause] it's hard to say, I didn't want to go into that, but I couldn't avoid it. I kept trying [pause] [laughs] but first they'd throw out the [formula for] circumference, she's [Rianne] just throwing them [the formulas] out, she throws them all out, then she finally gets it right, also, cause she hasn't memorized them, is that what's important? Or should you, this one you should be able to tell, there's no base or height in the problem [pause] this one, I could see where you could, but still, I couldn't, I could always keep it straight because r squared is an area that I knew, one r is just one, it's a direction.

The following is an excerpt of the lesson that Mark referred to in the previous passage:

Jamie: x is pi.
Mark: Yeah, so, so, 2 times pi, where this is the diameter, is equal to, the circumference around.
Jamie: Mmhmm.
Mark: Now, the area, you guys remember the formula for that? [pause]
Rianne: Base times height divided by 2.

Mark: That's for a triangle, that's right for a triangle.
Rianne: Pi times diameter.
Mark: That's the formula for the distance around.
Mollie: Oh, area.
Mark: Yeah.
Mollie: Area, is, area, times radius, equals, circle, I mean, pi times
Jamie: Radius squared.
Mark: Radius squared, right, what's the radius here, if the diameter's 2?
Mollie: 1.
Rianne: Radius squared times pi.
Mark: That's right, right.
Mollie: 1.
Mark: Right, so 1 squared? 1 times 1.
Mollie: 1, so it's going to be.
Mark: So the area's going to be pi.

On only a few occasions were the learners were encouraged to engage in reflection or critical thinking in order to construct mathematical understandings. The learners' weak conceptual understandings appear to have inhibited their ability to transfer the mathematical concepts covered in the class to new situations. The learners' own educational backgrounds and experiences appear to have prepared them for more traditional forms of mathematics instruction. The learners indicated that they often felt confused when the teachers abandoned a traditional step-by-step approach to instruction and introduced more investigative methods of learning mathematics. Rianne's comment highlights the learners' awareness of, and discomfort with, the preservice teachers' attempts to introduce exploratory teaching methods:

> Well, I understand that they, they want us to figure out things, you know, on our own, they're trying to do that, somehow, for what we're talking about, but I like it when you present it first when, you know, so you actually know later, instead of being asked later, cause when they question you, you're not sure where they're going, you know, they ask questions when we're kind of clueless what they want, and it makes it more complicated, trying to figure out what they want, yeah, yeah, and I don't like all that jumping around a lot, you know, and I understand they want to do this, but I like to see the themes, the categories, and then, when it comes to a problem, I like to see it go step by step, but they're jumping all around, because they could be talking about one thing and then it changes, and they're actually doing something else, and then it's not the same thing on the board, you see?

During a break between class sessions, Johnny joined a discussion with the preservice teachers about conceptual knowledge. His comments provide additional insight into both the learners' and the preservice teachers' struggles, and frustrations, with conceptual knowledge:

Toni: Well, it seems to me like they [the learners] were getting the arithmetic but they were really crunching on the concept behind it, and they, you know, they knew how to do the formulas.
Johnny: I don't like the whole concept thing, I never been learned, the concepts before.
Toni: Yeah, it seemed like all of you were kind of like bumping up against that.
Johnny: No one ever taught me the concepts; every math class I've ever been in has never the concepts thing going along with it.
Mark: What do you think of it? Does it hurt your brain?
Johnny: It sucks. [laughs]
Toni: Hurts your brain, yeah. [laughs]
Johnny: Maybe it's like something new, I usually, I've seen most of the stuff.
Toni: New stuff usually does hurt the brain. [laughs]
Mark: Does it give you a different way to think about it though?
Johnny: It might be just the squares that you guys use, I don't know what it is.
Toni: Yeah, uh, if we could think of
Johnny: I understand why you guys are doing it, it's just...
Toni: Maybe we could think of an alternative method that would
Johnny: It makes me feel like I just want to quit doing the problems.
Toni: Yeah, I hear you.
Johnny: Like I don't want to do this anymore.
Mark: Right.
Johnny: To learn the concepts, but there's some things that you just can't learn.
Toni: Restate, restate the concepts in a different way, is what I want to do.

The learners' weak understandings and poor retention may be explained by referring to Hiebert and Lefevre (1986) who note that a mathematical environment in which minimal attention is devoted to the development of conceptual knowledge creates an environment where learners are likely to forget previously acquired knowledge. Difficulty in recalling prior knowledge may occur when the algorithmic procedures or rules are presented in contexts devoid of meaning and isolated from other knowledge: Procedures that lack connections with conceptual knowledge may deteriorate quickly and are not reconstructible; they may be only partially remembered and combined with other subprocedures in inappropriate ways; they often are bound to the specific context in which they were learned and do not transfer easily to new situations; and they can be applied inappropriately without the benefit of a validating critic to check the reasonableness of the outcome (Hiebert and Lefevre, 1986, pp. 21-22).

Johnny

Beliefs

According to Johnny, mathematics was different than other subjects. He believed that, because of its complexity, mathematics required a skilled teacher who could show learners what to do. In contrast, he noted that it was easier to figure out what to do on one's own:

> You have to have the right teacher... in other subjects it's not as hard to know how to do stuff, I mean writing a sentence is completely different than writing out and figuring out a problem, an equation, yeah, because you know how to write a sentence, it's not that hard, just, "you are going to the park," period, and the other things you've got to figure out a lot more, it's more complicated.

Johnny believed that success in mathematics was a matter of mastering the procedures. It appears that he interpreted achievement in mathematics with the ability to successfully solve equations and complete procedural tasks:

> Math's easy as long as you know how to do it, you don't know how to do it, it's going to be awhile and then once you get to do it, then you'll know it and then you'll like to do it, it's kind of fun to do actually, math, probably, I'll have to do math, you know, I kind of like it.

Perhaps as a result of his limited exposure to conceptually-based teaching, Johnny believed that the most important reason for learning mathematics was "it's all your life you're going to have to add things, you know." He noted that doing taxes required knowing mathematics and "once you get into life, because there's a lot more in math than you think, you know." Johnny believed that he could learn mathematics better if he had a better teacher or if the subject were taught differently. He explained that a good mathematics teacher was personable, would help students out when they needed assistance, and "who's there for you whenever, like even if you're not even in their class, if you think they're a good math teacher, you go to them for help." He also noted that "a mathematics teacher should, always care, always care, that's it yeah, don't give up on a student because they don't know what they're doing half the time."

Self-Concept

Johnny attributed success to the fact that he considered himself to be "good at math." However, observations of Johnny, and conversations with the preservice teachers, revealed that his mathematics self-concept was rather low. The preservice teachers concluded that Johnny was reluctant to answer questions in class because he feared his misunderstandings would be revealed to the other learners as well as the preservice teachers. According to Johnny, he experienced difficulties with the mathematics courses he took at the mainstream high school because "the guy didn't put it in terms I could understand." He indicated that, if mathematics was "put in the right words," he could understand it. Johnny had previously encountered a variety of negative experiences with school mathematics. He elaborated on these experiences and how they shaped his conviction that:

> Math teachers should just always really care about their students, that's all I really want to say about that, because people just don't seem to care as much as they should, I mean if teachers want to teach, do it right, don't do it wrong,

> do it so they understand it, and they know how to do it, then they don't have to worry about it any more, and you know, maybe they'll more high school graduates, you know, they don't seem to care.

Learning Patterns

For the most part Johnny was an attentive and willing mathematics learner. However, his weak mathematical background often thwarted his attempts to fully participate in the planned mathematical activities and discussions. Of the five learners, Johnny was the student who was most likely to take notes and copy down the formulas and problems that the preservice teachers had written on the board. This was consistent with Johnny's statement that he learned mathematics best by watching and copying the teacher. It is also consistent with his assertion that he relied upon the teacher to affirm the correctness of his answers, "I usually just ask the teacher, did I do like, did I set this up right, yeah, [pause] if it's right [laughs]." Johnny's belief that mathematics was a body of facts to be memorized was evident in the way in which he responded positively to a step-by-step, formula-oriented approach to instruction. He often appeared to be quite reluctant to participate in exploratory activities and respond to open-ended questions. Johnny displayed a self-deprecating manner. For example, when Johnny was unable to solve a problem, he would often remark, "it's an easy problem" or "it's not that hard."

Mathematical Development

Because Johnny was often reticent during class, it was not always possible to ascertain his mathematical understandings during class discussions. However, Johnny was quite vocal during interviews. He demonstrated that he was able to recall a variety of mathematical facts and formulas but was unable to apply them with any consistency. Moreover, he did not appear to possess an understanding of the concepts underlying these facts. For example, during an early lesson Johnny mentioned the quadratic formula. However, he was unable to recall the formula or its application. Similarly, during the initial interview, Johnny explained that algebra, "would be like the x plus 3 equals 6 or something." Johnny recognized that, "I know basically, I know a lot, but I don't have a, don't know a lot of it well, I know how to do it, but I don't know it perfect."

Geometry and Spatial Sense

Johnny identified geometry as, "angles, obtuse, angles" and displayed little knowledge of elementary geometry concepts. For example, when asked to find the area of a rectangle, Johnny stated, "you add all sides, you either add or multiply both sides." Johnny's confusion about the formula for area persisted. For example, at the final interview, he explained his strategy for finding the area of his foot, "I'd think to trace my foot, probably [pause] and then [pause] find out the base and height, and, multiply, to find it." However, although Johnny was unable to correctly recall the formula for area, he appeared to have some understanding of the concept of area, indicating in class that the area was "all the area inside." Johnny displayed spatial sense during an early interview when he examined a rectangle and a parallelogram and noted, "They're both the same shape and like, that one's stretched out a little bit on the sides, this one's 14 and that's 12 so they're not exactly the same size but." Johnny noted that "If you had a piece of fabric that fit inside this one [the rectangle], if you stretched it, it would fit in here [the parallelogram]." Johnny was also able to triangulate polygons and recognize the connections between the number of triangles in a triangulation of a polygon and the fact that the interior angle measure increases by 180 as the number of sides increases by one. He was successfully able to utilize a one-square centimeter gridded transparency to measure his foot. He then noted that the area would change by a factor of four when he used a gridded transparency with sides half the original length (.5 cm). On a mathematics test Johnny recognized that he could find the area of a figure by breaking down the area into component parts. However, his inability to correctly recall area formulas prevented him from successfully utilizing this strategy to find area. He was unable to recognize that he could find the area of a trapezoid by dividing it into two triangles and a rectangle. Instead, he attempted to find the area of the trapezoid by turning it into three triangles. When Johnny was unable to use the three triangles to find the area of the trapezoid, he pursued a different strategy:

> You go like that, flip this one around, so it'd be over here [sketches a triangle on one side of the trapezoid to change it to a rectangle] and I'd say [pause] they're both 18 now, yeah [pause] and the side is 7 [pause] well, I've reduced this side, so it wouldn't be 18 any more [pause] hmmm, it wouldn't be 18.

Johnny struggled with the problem for several minutes, stating:

> I wouldn't know how I would do that [pause] I don't know, I know how to do it, I mean, all you gotta do is flip it around and do that, and then, you go ahead and do the base times height thing, it's not really, that hard, but the thing is, it's not to scale so, [pause]

Johnny was unable to utilize the given information to solve the problem. He indicated that he would need to measure in order to calculate the area of the trapezoid. Although he was unable to determine the area of the figure, Johnny provided some evidence that he understood the concept:

> 7, and you go 7, and then, if it was to scale, I could measure 7, with this, and go 7, 7, [measuring with fingers the distance of the 7] and maybe it would be 14, it would be 14, both sides would be 14, then it'd be 14 times 7 and that would be, the area, so you know I, I basically have the concept down.

Despite Johnny's procedural focus, he occasionally offered explanations which implied that he had developed some understanding of the underlying mathematical concepts. Johnny did not reiterate a formula to explain how he knew that any four-sided figure polygon would have internal angle measures of 360°. Instead, he offered an explanation which suggests that he did consider that there was a reason for his answer that went beyond applying a formula:

> Johnny: Because if you're to look one way, say you're facing the wall, and you turn completely around, that's going to be 360 degrees, and you're going to wind up facing the other wall, that's, I can't really explain it.
> Researcher: Does that have anything to do with it having 4 sides?
> Johnny: Yeah, cause each side is, you're going to turn 90 once, then 90 for each side, 360.

Number Sense

Throughout the course of the semester, Johnny was focused on both notation and procedures. As previously noted, he often resisted conceptual explanations and indicated that he preferred to be provided with formulas. For the most part, when he was able to recall and apply formulas, Johnny was able to solve routine calculations with ease. However, Johnny made several arithmetic errors over the course of the semester. Johnny's comment, from a classroom transcript, provides an example of his limited number sense, "I need to find out how to multiply that by that, I don't know how to multiply a quarter by 2." Because of his procedural focus, Johnny appeared to have difficulty ascertaining the reasonableness of his answers, as in this excerpt from the second interview session. When asked to estimate an answer for (800 x 99) + 801, Johnny resisted and requested time to calculate his answer. He worked on the problem and then displayed his answer:

> Researcher: You got 80,0001. [reads Johnny's answer]
> Johnny: Yeah.
> Researcher: Does that make sense in terms of the problem?
> Johnny: [looks at problem] Yeah, I don't know, I mean I don't really pay attention to how I learn it, I know it, how to do it, I don't know how to explain whether it makes sense or not, I just know that's the answer because that's how the math that's the way I figure things out.

Johnny made several comments in class which suggest that he recognized that he had difficulty with word problems. When working on word problems during a class session with Kevin, Johnny stated, "See, I don't know, I don't know, I can't set it up, that's all, that's the only problem I have." In an earlier class, Johnny had commented, "I'd rather just look at the numbers. I don't like story problems."

Functions and Graphing

Johnny demonstrated limited development of understandings of functions and graphing throughout the semester. Johnny's focus on the procedural aspects of mathematical problems appears to have inhibited his ability to transfer his mathematical understandings to novel situations.

Researcher: Which is line is steeper? [shows $y = 5x$ or $y = 2x + 7$] these are the equations for lines, which line would be steeper?
Johnny: [reads problem] Uh, I don't know, did we do this in class?
Researcher: You did a little bit out on that graph, similar to these, but again, this just has a little twist to it.
Johnny: Now these is two different problems though huh?

Johnny's low self-concept with regard to algebraic notation is displayed in the following transcript excerpt. It also demonstrates his procedural orientation and his limitations in arithmetic proficiency. When presented with "Solve: $3x + 4 = 9x - 5$" Johnny immediately stated that he did not know how to solve it. However, with encouragement, Johnny was successfully able to solve the problem:

Johnny: I know what you have to do, but I don't know how to do it.
Researcher: That's ok, do you want to just talk about what you have to do?
Johnny: Basically all you gotta do is get the x's on one side and once you do have them on one side, you can solve it pretty easily, that's, I need to find out how to do it.
Researcher: Ok.
Johnny: Can't you just like, you just go and put like minus 5x, can I write on this?
Researcher: Mmmhmm.
Johnny: You go like this, 9x, 9x [subtracts 9 from each side] like that, and cancel this one out and 6x [writing] uh, plus 4 equals minus 5 [continues problem] can I do that?
Researcher: Yes.
Johnny: [continues solving, gets $6x = 9$] You can't do that.
Researcher: No, no, that's ok, go ahead.
Johnny: [continues solving and gets $x = 1.5$]

Although Johnny appeared to be fairly proficient in solving simple equations, throughout the semester he exhibited difficulty in finding and setting up equations. For example, when Kevin gave the learners, "one number is 3 times another number, the sum of the numbers is 84, find the number", Johnny resisted having to find an equation:

Johnny: You know, I don't need to do this one.
Kevin: Why don't you try it?
Johnny: See the reason, I didn't say it was easy, I said that it was so easy it was hard.
Kevin: What do you mean it's hard?
Johnny: Because, like, once you get the, the equation on this, it's so easy, there's no point in even having, just the equation, if you can just look at it and solve it right then.

Johnny again displayed his procedural orientation during a lesson on graphing linear functions. As Mark and Toni led the learners in an activity on the life-size Cartesian axes, Johnny expressed confusion. He also revealed his preference for procedural thinking:

> Mark: Are you guys following this you think?
> Johnny: No.
> Mark: No?
> Johnny: I'm kind of understanding it.
> Toni: We're trying to 3x [pause] plus 7.
> Johnny: Is this basically what we're doing this thing here [points to equation on board y = 3x + 7]
> Mark: Yeah, that's exactly what we're doing, yeah, so they're plugging in some numbers for that, yeah, any number for x yeah, and that you get a value for y, and after you get like 3 points, you make the line, for that.
> Rianne: Oh, ok.
> Johnny: Just, this [referring to graph] is a harder way of doing that.

Johnny displayed limited understandings of function. When asked to read a word problem and determine if it was a function, Johnny responded, "I don't understand exactly what, I don't even remember what a function means." Similarly, Johnny was unable to determine whether or not the sequence 1/(2^n) was linear:

> Researcher: Is this sequence a linear sequence?
> Johnny: [pause]
> Researcher: Do you remember doing linear functions in class?
> Johnny: Not a whole lot.
> Researcher: Ok.
> Johnny: [pause] I would probably guess, yeah, it is, just because it's like, I mean it's all the same, it's all itself, you know what I mean?
> Researcher: The way it's changing
> Johnny: Yeah, it's changing by itself so, it means that any number you get, it's all even.
> Researcher: How you add ½ to ½ to get 1/4, then 1/4 to 1/4 to get 1/8?
> Johnny: Mhmm, you add up and you end up with the next[ ]
> Researcher: What made you say that?
> Johnny: I'm just guessing, you're just adding, there's a pattern, I have no idea whether it is or not, but, I'm saying
> Researcher: What is that pattern?
> Johnny: The pattern is, they're all repetitive, they're the same, yeah, if, if, you would've have like 1/3, f of 1, equals 1/3, f of 2, equals, 1/632 [laughs] I'd just be like, what is that, you wouldn't even know, but this one's so simple, it's so easy to find that, it seems, logical, to call it linear.
> Researcher: When you said f of n equals itself too, what did you mean by that?
> Johnny: It means just times every number by itself. ½, you don't times it but you add it by itself, ½ plus ½ is 1/4, 1/4 plus 1/4 is 1/8, you just add it by itself, you

> keep on going down and down, so, I would like to go f of n equals, 2 times, x or something like that.

Johnny was then asked to compare an equation (f(x)=3x+2), table of values (x=1,2,3,4 for f(x)=5x), and graph (f(x)=-2x-2) and determine if they were the same, or different, functions. These functions are denoted in the following quote as a, b, and c, respectively. According to Johnny, the functions differed from one another because of the presence of negative numbers in the solution set for one of the equations, "There are no negative numbers, in either one, a or b, there's negative numbers, only negative, yeah, not only negative, but a lot of negative numbers in c, since there's no negative numbers in a and b, they can't be any negative numbers in c if a and b were equal to c." Johnny's weak understandings of functions and graphing were further revealed in his explanation of why a graph of a function differed from an equation or table of values:

> I don't know [pause] I don't know, there's no equation on c, unless you wanted to make one, c by itself is just, a graph, you know what I mean, there's nothing there, man, and there's no way to figure out a problem, there's a way to figure out a problem for it probably, it's just that I don't know how do it, [pause] I don't know exactly how to explain it, it's hard to explain, [pause] you understand what I said though?

Rianne

Beliefs

Rianne shared Johnny's belief that success in mathematics was a matter of mastering a battery of procedures, "Once you can understand it, and find out a simple way to solve, the problems, when you do it, then that'll just, that'll give you some confidence to you know keep going on and doing it." According to Rianne, mathematics was different from other subjects because, "it's a lot of numbers, numbers, all the other subjects are like words." She shared a belief with the other learners that the most important reason for learning mathematics was because of its usefulness:

> Just in case you ever have to use it for whatever, you never really know, cause people don't really know unless you're going into that field, so, you may need to know, how to do it, how to solve problems [pause] um, dealing with money [laughs] and anything like that, yeah.

Based upon her own negative experiences with school mathematics, Rianne formulated a description of the characteristics of a good mathematics teacher. According to Rianne, a good mathematics teacher understands the way that an individual student learns mathematics and will identify for each learner the best way for him/her to approach a problem:

> Somebody who doesn't, somebody who can single you out and know um, the way that you do math, basically the best way that you can learn math, and they will form to that, and um, they'll help you out instead of forcing you to solve it that way that works best for them, for them to know how easy it is for you learn how to solve a problem.

Rianne revealed that she could do better in mathematics if it were taught differently. Rianne's uncertainty about whether a good mathematics teacher would show a learner many ways to look at a problem is compatible with her statements in which she expressed confusion at the preservice teachers' use of multiple methods and explanations.

Self-Concept

Rianne cited "luck" as playing a role in her mathematics achievement. She also indicated that when she got a bad grade in mathematics it was a result of not working hard enough and the teacher not liking her. Although Rianne believed that good grades in mathematics were a result of hard work, she also expressed a firm belief that success in mathematics was due to the teacher liking her. Rianne and Wendi were the only participants who associated teacher' attitudes towards learners with success or failure in mathematics. Similarly, they were the only two participants who attributed success or failure in mathematics to luck. Rianne's low self-concept appeared to be related to prior negative experiences with school mathematics. In a note to the preservice teachers, written after the first class session, Rianne stated:

> I've realized that I've forgotten a lot about mathematical ways to solve many of the problems that we had to work today. But it's ok, I'm not discouraged about it, I will learn and get better at whatever it is I will be taught. Math is my worst subject but I won't let that get to me, I am determined.

Rianne admitted that she was intimidated by Toni. It appears that, because of her own insecurities as a mathematics learner, Rianne was uncomfortable with Toni's attempts to coerce her to vocalize her mathematical understandings in front of the class. As a result, Rianne would skip class, or leave early. This missed time in the classroom may have adversely impacted Rianne's mathematics learning. Rianne's comments about teaching and learning mathematics provide some insight into her reticence:

> I'd say some of them do favor, um, and, some of them don't take the time out to really get to know me, and make you comfortable within yourself to know that you can do that problem, you know, if you know, if you really, something you never dealt with, you know, they don't you know, you just watch other people solve it for you and they don't, you know like this class is a good class, they help you solve it, you know, they really care whether you understand it, but other teachers, they just move on, even if you don't understand it, they'll just move on to a different area and just keep adding them on and you're

getting confused cause you're still trying solve the other one, and then they add on another problem, it really gets confusing.

Learning Patterns

Rianne was a rather independent learner. She brought her own calculator to class and would quietly try to work through the problems on her own. She spoke in a soft voice and her comments were often missed by the preservice teachers. There were several instances when Rianne appeared to be afraid to ask questions, speak up in class, or actively participate in open-ended activities. Rianne's learning patterns suggest that she was more comfortable working quietly within a closed-ended, procedurally-focused approach to mathematics instruction. For example, during an open-ended lesson on graphing when both the learners and the preservice teachers were becoming confused, the following exchange took place between Johnny and Rianne: Johnny: Plus 2, so why does everything seem so hard? Rianne: Cause people are doing it different. The following note, written by Rianne on one of her assignments handed in to the preservice teachers, is typical of the way in which she communicated information about herself as a learner to the teachers: "I tried as hard as I could to convert larger to bigger and vice versa and to understand and I caught on to the basics, but when we have to lay it all out, that's the difficult part. I know what I need to do thou so I'll be okay." Rianne revealed that she learned mathematics best by employing a variety of strategies. She carefully wrote down the procedures demonstrated by the preservice teachers and then employed these methods as she quietly worked through problems on her own. She also stated that she knew an answer to a mathematics problem was correct when she "double-checked" it. During interviews in which Rianne solved mathematical problems, she used a "double-checking" procedure to verify her answers. However, Rianne's rote method of double checking appeared to inhibit her ability to ascertain whether her answers made sense. Rather than enabling her to reexamine her strategies, this double-checking method merely resulted in her making the same procedural error twice. For example, when asked to explain why an incorrect answer appeared reasonable to her, Rianne responded, "That is the right answer, and how I know that? Because I double-checked it."

Mathematical Development

Although Rianne appeared to be a procedurally-focused learner, she was able to develop understandings with regard to properties of geometric objects, spatial sense, and function. However, this procedural focus may have inhibited the development of number sense. Her reluctance to ask questions in class may also have constrained her mathematical development. The lack of opportunity for practice and reflection also seem to have had a negative impact on Rianne's mathematical development. Although she made gains throughout the semester, she displayed an inconsistent ability both to recall procedures and to articulate her conceptual understandings.

Geometry and Spatial Sense

Rianne entered the course without any coursework in geometry. She was unable to discuss the topic in any depth and simply equated it to a school course, "I just think of geometry as like the next, actually, I don't really know how it goes, is it like algebra and then geometry?" At the onset of the semester, Rianne displayed a rather weak knowledge of elementary geometry concepts. For example, during an early interview, when asked to find the area of a rectangle, she responded, "add them up [the sides] and times that by 2." Rianne's misunderstanding of area persisted throughout the course. At the final interview, Rianne explained that she could find the area of her foot by either multiplying "length times width" or by "measuring around". During the same task, she suggested that doubling the size of the unit squares which were used to calculate area would double the numeric area. Rianne continued to have difficulty with the concept during a subsequent class session. On a mathematics test, Rianne recognized that she could find the shaded area of a figure by breaking the area down into component parts, "I know to find and add up the areas of each shape, then subtract the non-shaded area." Although Rianne was able to devise a strategy to solve the problem, she revealed that she could not recall the formulas needed to actually solve the problem. However, Rianne displayed her inability to extend this knowledge when she was unable to employ a similar strategy to find the area of a trapezoid. Rianne's attempts at solving many geometric tasks consisted of reflexive attempts at finding an appropriate procedure to apply in a given situation. However, when encouraged to reflect on the tasks, she was often able to solve them. For example, on a task in which the measures of the angles of a triangle were given as x, 2x, and 3x, Rianne immediately tried to apply the Pythagorean Theorem. Nevertheless, when forced to reflect on the task she was able to solve it flawlessly, identifying the need to set the angle measures equal to 180 degrees:

> Mark: Right, so if you had something like, um, this was x and this was 2x, and this was 3x, what would that be, how could you do that one?
> Rianne: We don't know x.
> Mark: What would you do then?
> Rianne: Use the Pythagorean.
> Mark: That's for the sides, these aren't the sides, they're the angles. What do they add up to?
> Johnny: 180.
> Mark: So you could write it out, right? x plus
> Rianne: Oh yeah.
> Mark: 2x plus 3x equal to 180.
> Rianne: And then 5x equal, 6x.
> Mark: You have 6x, yeah.
> Rianne: Divide it.
> Mark: Equals 180.
> Rianne: You have to divide.
> Mark: Right, so what do you get?

Rianne: 30.
Mark: Right, so this is 30, and what's that?
Rianne: It's 60.
Mark: Right.
Rianne: 90.

By the conclusion of the course, Rianne had become particularly adept at successfully completing a direct task. When asked to find the area of a given circle, for example, she was able to identify and apply the appropriate formula. Rianne's spatial sense was revealed when she examined a rectangle and a parallelogram and noted, "They're the same but then this one is just slanted." However, although Rianne was able to recall that the interior angle measure for a rectangle was 360°, she was unable to determine the total interior angle measure for a parallelogram, suggesting that she did not recognize the relationship between angle measure and number of sides of a polygon. Rianne's spatial sense did develop enough to allow her to obtain insight into some geometric results and ideas. She was quite proficient at identifying many different kinds of properties of polygons, including those associated with angle measures, lengths of sides, and area. This allowed her to make comparisons between polygons based on these properties. During a class discussion in which a rectangle with a base of 3 units and a height of 5 units was "slanted", Rianne conjectured that "when you slant it, it gets bigger?" However, after seeing different parallelograms with various slants, including one with a base angle almost zero, Rianne recognized that the area was approaching zero. Overall, Rianne seemed to develop an awareness of various properties of polygons as well as understandings of the spatial relationships between them. However, her ability to think about area was based almost exclusively on procedures. Given her few formal experiences with geometry, Rianne appeared to have built a solid base on which to develop further geometric understandings.

Number Sense

Rianne entered the course with limited ability to perform arithmetic calculations. Her overall number sense was also quite marginal. Early in the course she could not convert fractions to decimals or find the price of an object when it had been reduced by 25%. Throughout the semester, Rianne appeared to remain focused on procedures and notation. She also employed her double-checking method which involved incorrectly performing a multiplication procedure twice. Perhaps because of her procedural orientation, she had difficulty recognizing errors or ascertaining the reasonableness of an answer:

> Ok, it says you bought 49 cans of soup for 99 cents each, so for that, let's see I'd first, I would times the 49 by 99, and [does 9 times 9, writes a 1, carries the 8 - shows the Researcher] is that right, 81? [continues multiplying] Ok, let me double check, yeah, ok it came out to four thousand dollars, and 851 [she then explained that her answer made sense because she had double-checked it.]

Rianne was heavily dependent on her calculator which she brought to each class and interview session. When given the task (800 x 99) + 801, Rianne used a calculator to find the answer, but could not think of another method of determining an answer when asked. Similarly, Rianne relied upon her calculator to affirm the correctness of her answers:

Researcher: How about this? [.3(100)]
Rianne: I need a calculator for that.
Researcher: Can you do it in your head?
Rianne: No. [plugs into calculator and gets 30]
Researcher: If you didn't have a calculator, how might you solve that?
Rianne: [long pause] I don't know, I'm not sure.
Researcher: Does your answer make sense?
Rianne: Mmhmm.
Researcher: Why do you say that?
Rianne: Because that's the answer I got from the calculator.

Rianne's procedural focus and lack of number sense persisted throughout the course. At the end of the semester she stated that 872/1740 was equal to 2, "I divided the fraction and got 1.9, which rounded would be 2." It appears that Rianne found the value of the reciprocal of the given fraction. More importantly, she was unable to recognize the immediate inconsistency in her answer to a fraction with a denominator larger than the numerator. Rianne's limitations in arithmetic proficiency and number sense did not prevent her from exhibiting proportional reasoning. She was successful on most tasks, and was able to discuss applications of proportional reasoning in very clear terms. During a discussion on scale, proportion and ratio, Rianne recognized that, "I can do that in art, take something little and blow it."

Functions and Graphing

Rianne demonstrated some ability to create literal expressions, but struggled on tasks similar to the problem: *"There are six times as many students as professors at this university."* Although she was also able to solve a variety of linear equations, she struggled with quadratic equations. Rianne exhibited solid understandings of variable and correspondence in the context of the pizza problem. However, when introduced to functional notation (f(x) = 3x-7), Rianne asked, "Don't you add the x's up, but before you do that you do something else?" This remark suggests that Rianne interpreted the notation as an equation to be solved. It also provides another context in which Rianne searches for a procedure to handle a mathematical situation. She also struggled when asked to graph y = 3x+7. Although she was eventually successful at plotting points, Rianne never appeared to grasp the more global linear concepts that were being addressed, such as slope and intercept. She showed very little understanding of slope when encountering functional equations, and made comments that suggested that slope was dependent on what x-value was input into the function:

Researcher: This next one is a bit like what you've been doing in class. [shows next problem: Which is line is steeper? $y = 5x$ or $y = 2x + 7$] These are the equations for lines. Which line would be steeper?
Rianne: Most likely [pause] [points to $y = 2x + 7$]
Researcher: Why do you say that one?
Rianne: Because, anything, 2 times x plus 7, ooh, actually no, hmm, I don't understand this.
Researcher: Ok, what part of the equation shows how steep it is.
Rianne: The 5 here ($y = 5x$), and in this one the 7 ($y = 2x+7$).
Researcher: Of these 2 which line would be steeper?
Rianne: It depends on what x is.
Researcher: What do you mean?
Rianne: It would be different depending on what your x is. Definitely, 'cause this could, 5 times something, isn't that what y is?

Rianne demonstrated confusion between slope and the corresponding y-values used to think about slope. Her struggles with these concepts also prevented her from modeling or identifying rate of change in linear contexts. By the end of the year, Rianne was able to discuss slope in regard to rise and run in a graphic sense, but was still not able to relate this idea to an equation or points on the graph. From her work with the life-size Cartesian axes, Rianne developed a kinesthetic understanding of slope that involved moving up and over. However, this did not appear to be abstracted into a more general understanding that could transfer to numeric or equation contexts.

Mollie

Beliefs

On questionnaire items designed to probe Mollie's beliefs about mathematics, she checked "maybe" for each response, indicating that she was not certain about her beliefs. However, a portion of an initial interview is illustrative of Mollie's belief that mathematics was "kind of different" than other subjects because:

> For me it's harder to understand it. Just the subject itself that, that the numbers and everything, it's just harder to, it's easier for me to understand like pronouns and that kind of stuff than it is for me to understand like geometry and algebra.

Mollie indicated that mathematics was important because "you can't live without it". She also noted that she would use it throughout her life, "like for budgeting and stuff." Although observations of Mollie indicated that she was a strong mathematics student, she appeared to be much more interested in reading. According to Mollie, mathematics teachers were different from other teachers because, " some of them are real formal and stuff and they're uptight and the

English teachers seem to be more like outgoing and understanding." Mollie indicated that a good mathematics teacher was someone who could "get personal and they really like get into you and not just the math and everything." Her open, relaxed, and friendly relationship with the preservice teachers suggests that Mollie did not categorize them according to her definition of mathematics teachers. Instead, it appears that Mollie viewed the preservice teachers as being interested in the mathematics, but also concerned about the learners on a personal level.

Self-Concept

Mollie was an extremely bright and motivated mathematics student. Although she attributed success in mathematics to hard work, she did not consider herself to be "good at math." Like the other learners who participated in this study, Mollie had negative experiences with school mathematics. Mollie acknowledged that she had difficulties with mathematics. She indicated that, "I think I try my hardest and I try to understand it and everything but I get really frustrated sometimes, all the numbers and signs and letters."

Learning Patterns

Observations of Mollie's learning patterns reveal that she was an alert, enthusiastic, and highly motivated mathematics learner. She was extremely vocal and not at all shy about speaking up and voicing her opinions and ideas in class. When she wanted to know a "why" answer to a mathematics problem, she would persist in questioning the preservice teachers until she received a satisfactory response. Mollie's strong computational skills allowed her to quickly respond to the preservice teachers' questions, often to the consternation of the other learners who were slower to respond. Observations of Mollie's learning patterns revealed that she learned mathematics by watching and copying the other learners and working through the problems on her own. Mollie was always extremely aware of the other learners' strategies and solutions. Moreover, she did not hesitate to challenge the other learners' responses if she had questions. Classroom observations also confirmed that Mollie eagerly worked through problems independently. She shared the preservice teachers' belief that she knew a mathematics answer was correct when she rechecked it. Throughout the semester, Mollie displayed an ability to quickly work through calculations. Classroom observations confirm that Mollie enjoyed the challenges of investigative, open-ended instructional strategies. She did not appear to find multiple methods confusing and did not indicate a preference for closed problems.

Mathematical Development

Mollie often focused on notational and procedural aspects of the mathematics being discussed. However, there were several instances where she showed solid understandings of the underlying structure of a given mathematical procedure or concept. As a result of some of these positive experiences, it appeared that Mollie's self-concept in regard to mathematics changed. Early in the course she articulated self-doubts and negative feelings about mathematics and her ability to do mathematics, "Once I understand what I'm doing I like it a lot and I get through it pretty fast, but when I don't know what I'm doing it's real confusing and frustrating" and "I don't know, I think I try my hardest and I try to understand it and everything but I get really frustrated sometimes." Mollie had the most dominant student voice in the classroom throughout the course. Students would often turn to her for verification of their own ideas, and she was a student leader in generating ideas and strategies during class discussion. Late in the semester Mollie was asked to solve a geometry task. Although she did not have an immediate strategy, her prior success and emerging confidence prompted the following comments:

> Mark: Here's one for you. [gives Mollie a 12-sided figure]
> Mollie: Huh? I couldn't do this.
> Mark: Sure you can, do you want to try this one first?
> Mollie: [laughs] You, I don't even think I could do that one, [pause] let me see [studies figure].

Her attitude of "if I can't do, then no one can" indicates a very positive self-perception of her mathematical abilities. Clearly, by this time, Mollie felt that she was the top student in the course.

Geometry and Spatial Sense

Entering the course, Mollie had little experience with or knowledge of geometry. Evidence of this can be found during an early interview, in which she was presented with an elementary geometry task:

> Researcher: Ok, this is the last one [shows diagram with rectangle with length 25 and width 12]. Can you tell me how you would find the area of that figure?
> Mollie: [long pause] No, I have no idea.
> Researcher: Ok, that's alright. Have you had any geometry?
> Mollie: Well I did, but I didn't understand any of it.

However, by the end of the semester, Mollie was able to investigate areas by triangulating given polygons and was skilled at approximating area by using gridded transparencies. She also become fluent with area formulas for rectangles and triangles, and had a tenuous hold on the formula for the area of a circle. Evidence on her sense of the processes of finding area indicates that she learned these skills with some degree of understanding. She displayed spatial sense when she noted that a parallelogram was "a rectangle that's been sort of slanted." Three days before this, during an in-class activity in which regular n-gons inside a fixed circle were being observed, she noticed that the limit of the area of the n-gons would be the area of the circle. The teacher then produced a graph of number of sides vs. area for several different n-gons. However, despite her observations and the presence of the graph, she continued to demand to use formulas to find area:

> Mark: Yeah, up somewhere over here, is it going to be, let's take a guess, will it be like over here? Or somewhere over here?
> Mollie: Over here.
> Mark: Over here?
> Mollie: Well
> Mark: What do you think Rianne?
> Rianne: Over there.
> Johnny: It would be easier if you knew the area of the circle.
> Mollie: If you calculated, if you just did it, if you just figured it out.
> Mark: Maybe you could do that then?
> Mollie: Well, tell me the formula and I'll do it, what's the formula?

Mollie displayed knowledge of special angles (90, 180 and 360 degrees). An activity in which corners of a triangle were placed next to each other to form a line, illustrating the fact that the interior angle measure of a triangle is 180 degrees, seemed to have a positive impact on Mollie's understanding of this fact. However, her belief in this fact was still tenuous. After a discussion involving similar triangles of different sizes Mollie asked, "is it always going to be like that, is there ever a time when it, when your angles don't equal 180 degrees?" After using a protractor to measure some given triangles, she became convinced. In the next class session she was able to immediately state that there were 180 degrees in a triangle, and then applied this knowledge to find the missing angle measure of a given triangle. Mollie then illustrated additional facility with triangles. When given an equilateral triangle, Mollie was able to deduce that the angle measures would be the same and apply her knowledge of the angle measurement of a triangle to calculate each angle as 60 degrees. She also stated that each side would be the same length. During a later interview, Mollie illustrated how far she had developed in regard to an understanding of area. Given a trapezoid of height 7 and bases 12 and 18, Mollie divided the area into two triangles and a rectangle. When asked to find the area, she said, "Do you want me to do the math?" She then calculated the area of the rectangle (7x12), and then computed 18-12=6 to get the combined length of the bases of the two triangles. She then computed (1/2)*6=3 and 3x7=21 to get the combined area of the two triangles, which she then added to the area of the rectangle to obtain the solution. This problem solving process indicates an ability to break down area into component parts, knowledge and proper use of area formulas, and solid problem solving skills. Over the course of the year, Mollie made procedural advances in the area of geometry in regard to angle measure and area. She also displayed understandings of the area process and was able to use spatial sense to compare and contrast geometric figures. However, evidence was found late in the year to indicate that Mollie may not have made solid learning gains in some geometric aspects that led to retention. During a late year review before the test on various polygons, she seemed to remember very few terms besides "isosceles". More importantly, during the next class session, she did not recall the derivation of pi, stating that "it's some number somebody made up." After some prompting, she recalled an activity that occurred just two weeks earlier that measured various circles to reconstruct the meaning of pi as the ratio of circumference to diameter. Despite these lapses, her understandings of area appeared to have remained strong by the end of the course.

Number Sense

Mollie was heavily focused on notation when performing arithmetic operations on fractions and decimals. When encountering addition and subtraction tasks, she expressed a need to write fractions "up and down" so that they could more easily be converted into fractions with common denominators. She was fluent in procedures on fractions, indicating the process of multiplying fractions was to "multiply across." Mollie expressed a resistance to thinking about fractions in more conceptual realms. During the second class session in which Toni introduced an array model for fractions, Mollie seemed to follow the ideas being discussed, but indicated that she preferred to find the least common denominator and continued to problem solve this way, "This is how I learned it, like this, you do the common denominator, the lowest common denominator which is 6, and then you go 2 plus, times 3 is 6 and then 1 times 3 is 3." In the next class session she reiterated her resistance to the array model. Her inability to work with fractions outside of procedural realms had consequences on her ability to perform other tasks. She had difficulty changing a fraction into a decimal, illustrating a lack of awareness of part-whole concepts and a reliance on procedures. Later, she correctly stated the procedure as "top divided by bottom", but continued to attach little meaning to the procedure. Two days later Mollie once again made a procedural error when converting 3/10 into a decimal. Mollie's procedural focus on number was also manifested in her general approach to arithmetic tasks. When encountering the penny doubling task, she stated that the "only way" to determine whether or not to choose the pennies on the chessboard would be to perform the calculations. In two that could easily be performed with estimation, Mollie chose to calculate. The following task, which involved a calculation of the price when 25% was deducted from the price of six 45-cent candy bars, provides a typical illustration of the manner in which both Mollie and Rianne tended to jump into an arithmetic task by applying procedures:

> Kevin: Right, this is basically, you put 2 there right? So it's going to be 270 [writes 2(3) times 45 = 2.70]. Now, what if, when he buys 6, what if you get 25% off? What are you going to do then? [long pause]
>
> Rianne: Divide by .25? Or multiply?
>
> Kevin: What, ok, you want to find, what is it you want to find? The number
>
> Mollie: Uh, a fraction of 270, so would you do 270 divided by?

Kevin: No, let's slow down for a second, what do you want to find? What's the point of this problem? You want to find out, if you buy 6 and he gets 25% off, you want to find out how much it costs, right?
Mollie: Yeah.
Kevin: Ok, now what do you know?
Mollie: The total cost is 270, that's how much the candy bars are going to cost, and then 25% off of 270, right?
Kevin: Now what if, what if it didn't matter how many you bought, how would you write an expression for that?
Mollie: Oooh!
Kevin: Do you know?
Mollie: 25 percent of n times point 45.

The final portion of this segment suggests that, when forced to reflect on the task, Mollie was quite successful at thinking deeply about the arithmetic processes involved, as she was able to generalize this computation into algebraic realms. But the beginning of this segment clearly indicates her tendency to initially approach problem solving situations in a procedural fashion. Mollie continued to struggle with fractions and decimals. She stated that 1/2 was less than .3 because, "I turned them all into decimals." An inability to properly convert a fraction to a decimal may have led to this confusion. However, a reliance on procedure and limited number sense did not allow Mollie to think about 1/2 as .5 and quickly compare the magnitudes of these numbers. However, Mollie showed good use of proportional reasoning on numerous occasions. For example, using the fact that the football field on a campus map was 1.5 inches, Mollie was able to deduce the scale factor for the map. Mollie showed very consistent and spontaneous use of proportional reasoning throughout the course. Overall, however, it appears that the instruction did little to influence Mollie's procedural tendencies when thinking about number and associated arithmetic operations. This is particularly true with regard to operations on fractions and decimals as well as tasks with large numbers.

Functions and Graphing

Mollie entered the course with minimal understandings of functions and graphing, in part due to bad experiences with past courses, "I did good in math and stuff but then like in 8th grade and 9th grade I was in prealgebra and I didn't, I understood the stuff. But I didn't go to class enough to get through it, so." Over the course over the semester, Mollie made considerable gains in her understandings and abilities associated with various algebraic topics. She quickly mastered the ability to generate literal expressions given statements such as "3 less than twice a number", and she used this ability to generate equations in various problem solving situations throughout the course. Although Mollie was generally successful in solving linear equations with a single variable term, she was less successful in solving linear equations which contained variables on both sides of the equation, a phenomenon Filloy and Rojano (1984) have termed the "algebraic divide." Mollie demonstrated understandings of functions, especially understanding the correspondence relationship between x- and y-values. However, her understandings of the connections between equations and graphs, slope, variable, and covariance were less well-formed:

> Mark: I just, I wanted to, this is something I want you to think of for a second. If y = 2x+1 is one line, now we haven't graphed anything like that, have we? And y = 2x-5, do you think they'd be parallel or not, what do you think?
> Mollie: They'd be parallel?
> Mark: Why?
> Toni: Why do you think?
> Mollie: Cause they're both, you're times-ing both by 2, times the x by 2.
> Mark: So what does that say about how it would be graphed?
> Mollie: That you'd be going 2 over.
> Mark: 2 over.
> Mollie: And then 5.
> Mark: 1 over and 2 up?
> Mollie: Yeah, 1 over and 2 up.

Mollie's tendency to find slope by dividing the constant term by the coefficient of x stayed with her throughout the year. However, at times she would also interpret slope using the coefficient of x in a more appropriate manner. The following discussion about the function f(x)=2x occurred during a class session, in which Mollie made use of the big coordinate axes taped to the floor:

> Mollie: It's like, if you're on, if you're on 3 . . . then it would be 6, 4, right?
> Toni: If you're on 3, 4, ok?
> Mollie: And you do x times 2, it'd be 6,2, right? [long pause] Yeah?
> Toni: Ok, and so [Mollie continued to state points as (y,x) rather than (x,y), something that apparently went unnoticed by Toni]
> Mollie: So that's when you take the x and you times it by 2.
> Toni: So how do you relate that to, to what Jamie was saying about
> Mollie: How they're connected?
> Toni: The x-axis being called x, who can relate that?
> Mollie: The x-axis is the whole line, it's just
> Toni: And they're all x's
> Mollie: Yeah.
> Toni: On it?
> Mollie: No matter what number you're on, x.
> Toni: Ok, so, 2x would still be on the, would be, would be
> Mollie: Yeah, it would just be whatever your x is times 2

This discussion illustrates Mollie's solid understanding of correspondence. However, in a discussion of the function y=3x+7 a few minutes later, she illustrates a more tenuous hold on the idea of slope:

> Toni: What would be the slope?
> Mollie: The 7?
> Wendi: The 7, 'cause the 7 makes the difference in whether it increases or
> Mollie: No, no, no, no, no. It would be the 3, it's what you times x with.

Jamie: You keep saying y equals, all this equals to that one line, going up?
Mollie: You're just trying to find out what x and y are.
Toni: Right now, all I want, what I want to find out
Mollie: Forget the parallel.
Toni: Is what's the slope?
Mollie: Slope, 3, it's what you times the x by, and it makes it get bigger and smaller, makes the slope increase and decrease.

Overall, Mollie's ability to appropriately analyze slope using graphs, tables of values, and story contexts suggests that her understandings of the meaning of slope were driven somewhat by functional notation, but that she had a sense of slope as a constant rate of increase. Mollie's work with graphing calculators led to increased understandings as well as confusions. When working the starburst activity, she made important discoveries about slope and y-intercept. She was able to connect the slopes of various lines to the coefficients of x in the associated equations, and she also was able to move the starburst up the y-axis 7 units by adding 7 to each equation. At times the graphing calculator seemed to confuse Mollie. When using TRACE to find the y-intercept, Mollie was led to believe that the function did not have a y-intercept because the graphing calculator would not TRACE exactly to $x=0$. Mollie seemed to continue to struggle with algebraic notation late in the year. During a year-end review she stated that $y=3x-2$ would have a slope of "3 over and up 2", and then participated in a discussion that revealed extreme difficulties in numerically interpreting fractional slopes. As in the case of her understandings of number concepts, it appears that Mollie had difficulty understanding the structural ideas behind certain symbolic notations.

Jamie

Beliefs

Jamie was the only one of the learners who was able to articulate a metaphor for mathematics, "Just like this big puzzle [pause] and you can figure it out and that's what I like about math, it just drains your brain, you work on a problem, and figuring it out." During an initial interview, Jamie described a utilitarian view of mathematics. According to Jamie the most important reason for learning mathematics was, "We're going to use it through our lives, it's very important, math is just in everything, like science, everywhere in the world we look we find math, and we're going to need it through our lives." Jamie suggested that mathematics teachers should get to know their students personally, including understanding how "what's going on in their lives can affect them," have patience, and care about their students, "that's very important." She identified the characteristics of a good mathematics teacher as, "Someone who's patient and really cares about the students and someone who can get the students to want to do the math because they want to - you can't, they won't do it for the teacher if they don't want to do it themselves."

Self-Concept

Jamie attributed success in mathematics to the fact that she considered herself to be "good at math." She enjoyed mathematics, stating, "I basically like math, I like to work problems." Her enjoyment of mathematics was evident to the preservice teachers who were impressed with the depth of Jamie's questions. However, Jamie admitted that she had difficulty staying focused and, because of that, required "a lot of one-on-one." This comment is consistent with her belief that success in mathematics did not require an inherent talent and that she could learn mathematics better if she tried harder. Jamie had negative experiences with algebra in the mainstream high school, stating, "That guy [the teacher] basically made it a bad experience for me, but I think, I know I could do it, maybe if I had some one-on-one, a different type of teacher."

Learning Patterns

Despite her numerous absences due to legal problems, Jamie was an extremely willing and attentive mathematics learner. Her concentration and participation grew over the course of the semester. At the onset of the study, Jamie would try to deflect the focus of the lesson toward a discussion of movies and rock stars. However, as the semester progressed, she became more focused on the mathematics and continued to raise intuitive questions. While Jamie's weak mathematical background and inconsistent understandings often thwarted her attempts to fully participate in the planned mathematical activities and discussions, she remained an enthusiastic learner. Subsequent observations and conversations with Jamie confirm that she believed learning mathematics was interesting and helped her think better. Jamie employed a variety of strategies in the classroom. Unlike some of the learners, she appeared to enjoy and was not confused by multiple approaches to problem solving. When she solved problems, she used a variety of strategies to verify her answers, including looking at a problem and recognizing that her answer was not correct. Although Jamie indicated that she utilized multiple methods to verify her answers, she and Wendi were the only two participants who indicated that an answer was correct if "it just looks right."

Mathematical Development

Jamie stated that she enjoyed the intellectual challenges of mathematics, "That's what I like about math, it drains your brain." She displayed both procedural and conceptual gains throughout the course. These gains were primarily in the area of geometry, as she continued to struggle with numeric ideas and algebraic symbolisms. Her academic struggles were compounded by her personal problems which disrupted her course attendance during the final three weeks of the semester. Jamie entered the course with a low mathematical self-esteem. When describing her ninth grade algebra teacher she stated, "He basically cared about the kids who were good in math and ignored the rest of us." When addressing mathematical tasks Jamie displayed a mind set that focused on notation and procedure. However, this did not stifle her inquisitiveness and need to understand the underlying structure of many of the mathematical topics that were discussed throughout the course. For example, she insisted on knowing where "3.14 comes from" in an area formula, why the x- and y-axes were labeled the way they were, and she questioned whether or not "bigger circles" would have more than 360 degrees in a rotation. Her question about the measure of an interior angle of an octagon led Brad into an entire lesson on this topic for various polygons.

Geometry and Spatial Sense

Jamie entered the course with virtually no background in geometry but made great strides in her knowledge throughout the course. She quickly identified special properties of angles, and used her knowledge of the fact that a line contains 180 degrees to finish the justification for the interior angle measure of a triangle. She was also especially quick to learn the meaning of various geometric terms, such as "acute" and "hexagon." Jamie made the most progress in the development of spatial sense with respect to area of polygons. She displayed solid understandings of the process of triangulation, and was able to use these understandings in making sense of the interior angle measure formula for polygons. Jamie was also able to apply these understandings of triangulation to polygons that were not convex. Jamie initially calculated the area of a rectangle by adding the base and height, and also saw no real connection between a rectangle and parallelogram with the same base and height. Although, in the following passage, she makes inappropriate statements about right angles, she appropriately applies and illustrates a conceptual understanding of the interior angle formula:

> Researcher: Can you say anything about the angles?
> Jamie: The angles? Um, they all have 90 degree angles, they're all 360.
> Researcher: Ok, how do you know that?
> Jamie: Well, I think of, the equation that we made up, that 4 sides, like, minus 2 [pause] um, 4 sides minus 2, 'cause I think of like, triangles, like that, where 2 triangles, so it's like 4 minus 2, times 180.
> Researcher: Right, mmhmm, good, would that be the same or different for this [the parallelogram]?
> Jamie: Mmm, it'd be the same, because that's 4 sides, and you can still split that up into 2 triangles.
> MI: You said they all have 90 degrees, is that, is that what you said a while ago?
> Jamie: Like, how they showed us how to do that [marks two 90 degree angles with pencil on the rectangle] like that.
> MI: What about the other figure [parallelogram], do they have 90 degrees too?
> Jamie: Um, yeah, it's a total of 360, they're all going to be 90 degrees, 90 degree angles.

When a similar discussion of rectangles and parallelograms arose, Jamie gave more appropriate attention to the individual angle measures:

> Toni: Here's my 3 by 5 [rectangle], and here I have, a 90 degree angle [pointing to an angle on the rectangle], and here I have a less than 90 degree angle [pointing to an angle on the parallelogram].
> Jamie: It's still all going to be 360.
> Toni: We'll just name that theta and we'll say it's less than 90 degrees.
> Jamie: It's still going to be, it's closer, these here are bigger than 90 degrees. These two here, are less than, so it's still going to be 360.
> After a discussion of the possibility of the two polygons having different heights, Jamie reiterates her claim:
> Jamie: Still, all those are going to equal 360 Those 2 are obtuse, and these ones are acute, but they're still, all going to equal 360.
> Jamie then turns her attention to the heights of the polygons:
> Toni: Alright, what's?
> Johnny: I'm not sure they're the same size though.
> Toni: Ok, well let's take a look at this. Do you think it would be the same?
> Jamie: But I mean the middle one, I mean if you took it and slanted it, it wouldn't be 3 inches any more, the corners [pause] so I guess the middle wouldn't be 3 inches, the area wouldn't be, the lines.
> Toni: The lines can still change. What if my, like Jamie just said, what if this was still 3 inches, this was still 5 inches, so it's, all these were the same, am I getting closer to this line?
> Jamie: That's changing.
> Toni: So this, this number here is not going to be 3 inches.
> Jamie: They all have to be 3s and 5s.

It appears that Jamie utilized her developing spatial sense to make connections between the individual and total angle measures of the rectangle and parallelogram, and also made connections between the slope of the side of the parallelogram and its relationship to area. Her understanding and procedural abilities with area were confirmed when she successfully used gridded transparencies to measure her foot, and then noted that the area would change by a factor of 4 if she used gridded paper with sides half the original length. She later broke down a trapezoid into a rectangle and two triangles to calculate the area, obtaining an inaccurate answer because of an arithmetic error. Jamie seemed to struggle with measurement concepts in relation to circles, perhaps because of her inability to fully understand angle rotation or her lack of understanding of pi. In any case, this caused problems for Jamie in relation to measurements on circles. Overall, Jamie displayed solid understandings of measurement concepts and spatial sense throughout the semester. Given her relatively little experience with geometry prior to the course, it appears that Jamie gained a considerable amount of insight from the geometric portions of the course.

Number Sense

Jamie approached most arithmetic tasks and contexts in a procedural manner. She performed estimation tasks by calculating but when prompted, was able to use estimation. Jamie often made arithmetic errors and relied on the use of the calculator quite heavily. Very early in the year, when faced with the task of adding 1/3 and 1/4, Jamie stated, "You have to get the bottom to be 12." She later successfully described the procedure for dividing fractions. When encountered with an array model for fractions, she saw no point in thinking about fractions this way. In the next class, she made the following statement in regard to this conceptual model, "I don't understand what you're doing when you do it visually." Clearly, Jamie preferred to work with fractions in purely procedural realms. As in the case of Mollie, this procedural focus had consequences on her ability to perform related computations. After a lengthy discussion of converting stock shares into percentages, Jamie stated, "I don't know how to change fractions to decimals." It seems that Jamie focused only on learning a method of converting shares to percentages and did not realize the more general applicability of this skill. Although she was able to accurately make conversions in the following class session, her thinking was still focused on the performance of algorithms. Jamie's lack of understanding of fractions and focus on procedures prevented her from performing a skill closely related to one previously encountered. Jamie's struggles with numbers were not restricted to fractions, as her lack of understanding of decimals prevented her from successfully multiplying .3 x 100.0, stating "Is it 2 decimal places?" In the middle of the course she also asked the following question with very little confidence, "When you're working with negative numbers, when you times 2 negative numbers, it's a positive?" However, despite her lack of number sense associated with fractions and decimals, Jamie showed procedural and conceptual understandings of ratio and proportion:

Kevin: Who can tell us about ratio?
Rianne: Ratio is a comparison, like chairs to tables, boys to girls.
Kevin: Ok.
Jamie: 6 chairs to 3 tables.
Kevin: Ok, so it's like a ratio of number of guys to girls.
Rianne: Or girls to guys.
Jamie: But we went way in depth more than that.
Kevin: Right.
Jamie: We had these big wrenches like this and then we looked at little wrenches like this and you had to figure out how we could get that big size down to the exact little size was what it was.
Rianne: You divide it.
Kevin: Right.
Jamie: Divide the little number and you get that like.
Kevin: So the concept of a ratio is comparing 2 things right, using like the quantity of them or the length, you could do that, you could probably do it for weight, for different things like that, um, now what do we know about scale?

Jamie: Wasn't, maybe that, scale, I don't know, like, what I just said would be scale.

Overall, Jamie continued to think about numbers and arithmetic tasks in procedural realms, showed little understanding of fraction and decimals, but on several occasions demonstrated solid proportional reasoning ability.

Functions and Graphing

Jamie exhibited limited development of understandings of functions and graphing and struggled with standard algebraic notations. She had great difficulty constructing literal expressions from phrases such as "2 less than 3 times a number" early in the year. Throughout the year she was not able to solve equations such as $10 + 2x = 24$, and she initially could not construct an equation to model a linear context. Her in-class experience with the pizza problem generated some understandings of modeling, as by the middle of the year she was able to model a linear context, but stated that "It's kind of hard to think without referencing it to pizza." Jamie also showed great difficulty solving linear equations which were not in a proportion context. The latter were usually written as fractions and required cross-multiplication to solve. The pizza problem illustrates Jamie's tendencies to prefer to think in procedural realms in the context of functions. When the problem was being introduced, Jamie appeared interested and involved in the context. When told that she would get $3 for each piece of pizza she ate, Jamie stated, "I'd make sure I ate a whole pizza!" As the discussion ensued, she made several insightful comments and created a solution strategy to a contextualized linear equation task:

Toni: Here's my pizza, it's still the same price. So let's take $7 to get in, what if we had $29 when we walked out? [long pause] Let me finish that question, if we had $29 walking out, how many slices of pizza did we eat in the party? [long pause]
Jamie: 12?
Toni: And how did you get that?
Jamie: I added the 7 and divided it by 3.
Toni: Ok, ok, [pause] you added the 7 to?
Jamie: 29.
Toni: The 29, ok.

However, her problem solving strategy was done without the use of algebraic symbols, and she continued to struggle in this domain throughout the course. Perhaps Jamie realized this, as just after her solution she and Rianne made the following request:

Rianne: What about having all the pizza and all that, can you just put like a problem that we can do up there.
Jamie: Yeah, just, just do a regular problem. Like what we tried to do on that, I didn't unders - and I

Rianne: Without, without, like the whole, you know, party theme for it.
Brad: You just want to solve it?
Toni: You just want to see solving a problem with an equation.
Jamie: Like there's one in there like minus, or, what was it?

The class then returned to literal expressions, with which Jamie continued to struggle. Brad then decided to introduce functional notation in a factory context, where x represented the output of the factory. This again brought resistance from Jamie:

Jamie: But I don't understand that.
Toni: What comes out?
Jamie: Because if x is what comes out
Toni: No, f of x is what comes out.
Jamie: Oh.

However, later Jamie understand x to be "whatever number", and also mentioned that x would be "slices of pizza" in the pizza problem context. Her general understandings of functions were heavily related to the pizza problem context, but she did make some generalizations in regard to variable and correspondence. This development continued as Jamie was introduced more deeply to linear functions. In an activity that utilized the life-size Cartesian axes, Melissa once again used the pizza problem function, which was written as $a = 3p\text{-}7$, where p is the number of pieces eaten and a is the amount of money earned. Melissa then assigned the students various values of p, had them calculate their corresponding values of a, and then stand on their respective points. After Johnny noted that "it makes a line", Jamie asked, "I that because we all had the same equation?" It appears that Jamie was beginning to make strides into general understandings of functional correspondence. During an activity that investigated the graphs of functions of the form y=x+k, Jamie noted that the "number you're adding to it changes" but never appeared to connect this to the graphic or numeric aspects of the function. However, she later made explicit connections between the graphs and equations of functions of the form y = 2x+k in regard to the y-intercept. Jamie also continued to have trouble with symbolic notations, as she plotted points associated with the function x=y+3 when encountering y=x+3. In the next class session she showed signs of beginning to understand the underlying concept of slope:

Jamie: Ok, so, we have 2x [pause] plus 7 and then 2x plus 3.
Toni: And, and so what were you saying for that?
Jamie: So then they would always be in parallel, because
Toni: Because?
Jamie: Because the 2 was always the same.

Later, she showed an understanding of intercept, but still had unstable notions of slope. Here the discussion is about the function y = 2x, and Jamie is addressing specific points:

Jamie: It would be 2 times 1, I would think.
Toni: So, it, why don't you show me over there? [points to life-size graph] Why don't a couple of you show me there?
Jamie: [goes to floor graph] Ok, so we're on the x line, right?
Toni: We're on the x line.
Jamie: Because we're on the x line, and it's this (y = 2x), this line is 1, right?
Toni: Ok.
Jamie: 2 times that would be 2, so you go over 2.
Toni: Ok, 2 [pause] If it, ok, if you're starting out with x and said it was 0,
Mollie: No, no [goes to floor graph]
Toni: x was 0, um.
Mollie: It's like, if you're on, if you're on 3
Rianne: Yeah, it's 0.
Mollie: It's like 3, 4, ok, and you times the x by
Jamie: [sitting down] Well, then I just don't get it then.

By year's end it was clear that Jamie had never learned or had forgotten most of the algebraic ideas discussed during the course. At the end-of-year review, when asked for an example of a function, Jamie stated, "that pizza one." However, she had trouble modeling an equation for this now very familiar context, and she did not seem to be able to talk at all about general linear properties such as slope and intercepts. Her limited understandings were confirmed during the final interview. After being shown the first four terms of the sequence 1/(2^n), she calculated f(10) by adding 2 to the denominator six times, obtaining 1/28. Although she later stated that "you are actually times-ing by 2", she was never able to orally or notationally describe the function adequately: "I don't know what a linear sequence is. I don't remember doing linear functions." Her inability to compare equation (f(x)=3x+2), table of values (x=1,2,3,4 for f(x)=5x), and graph (f(x)=-2x-2) (denoted in the discussion below as a, b, and c, respectively) clearly indicates her shallow development of algebraic ideas and notations.

Jamie: [pause, studies page] They all represent the same function.
Researcher: Ok, why do you say so?
Jamie: Cause, all of them come out to be the same answer [pause] I mean all of them are, um [pause] the same, just written differently.
Researcher: How do you know that?
Jamie: Because, ok, you, you're given the equation, the function, and that's the written function, but then x over that, xy scale thing is um, [pause] is the same, it's just, writing it, differently, it's putting in the variable. Oh, well then I guess it would be different, 'cause you're putting in the unknown for x [pause]. You're giving x a number, so it, I mean you're making x a number.

Here Jamie is stating that the table of values is different because it involves specific points that have been evaluated, whereas the graph and equation do not:

Jamie: Yeah, so I mean I guess that would be different, but [pause] just, that the function doesn't have a number yet, it's just a x, and it's an unknown, and then when you do b, you're giving it, the, you're giving it a known number, so it's not unknown.

Researcher: Ok, are you allowed to have, known numbers in that function (the equation in a)?

Jamie: Yeah! There's 3, and 2, those numbers.

Researcher: Can x ever be a known number?

Jamie: Yeah! It is in b. You're making it that way, and c (the graph) is b (the table of values), you're just putting it in.

Researcher: How are b and c related?

Jamie: b is, is c. You're just putting the dots on the graph. [Here, Jamie has made a connection between graphs and tables of values in regard to points, but she has yet to do so for equations].

Researcher: How do you know they're the same dots?

Jamie: Oh, I haven't looked yet but [studies graph] Oh, they're not, cause this is minus 1 (the x-intercept of the graph in c), those are negative numbers, and there's no negative numbers in b.

Researcher: What if this part of the graph was covered? [covers area of graph below x-axis with hand]

Jamie: Mmmm, well [pause] mmmm. It's still not right, 'cause this is going in the negative direction, and, these are asking for positive numbers and positive numbers are over here. [points to quadrant]

There appears to be a connection between graphs and tables of values in her ability to recognize that the two functions would have no points in common. However, the difficulties when working with equations persisted, and Jamie never focused on anything besides points in her analysis, such as the properties of slope and intercept. Jamie left the course with very little understanding of functional ideas or sense of algebraic notation.

Wendi

Beliefs

Wendi articulated a belief that mathematics was different from other subjects because "it requires more concentration." She expressed a utilitarian view that it was important to learn mathematics because of its usefulness, "so I can get by in life." Wendi indicated that mathematics was going to be very important in her future "because, I have to know how to cook, you need to know math for that, you have to know it for budgeting, groceries, rent, things like that." She indicated that she didn't want to learn mathematics and it appears as if her own personal motivations for learning mathematics were primarily extrinsic. Wendi revealed that she enrolled at the mathematics class at the suggestion of her counselor. According to Wendi a good mathematics teacher was, "Someone who understands their students, who can, like, be there, you know, who can like understand that they're going through stuff and like give them a little leeway and not put more stress on them.."

Self-Concept

Wendi had endured a variety of negative experiences with school mathematics. She stated that she disliked mathematics "just cause nobody's ever helped me like understand it." She acknowledged that she could learn mathematics better if she tried harder. Wendi also revealed that, as a result of "horrible" experiences with school mathematics, she believed she was "not very good" at mathematics. She expressed a belief that success in mathematics was due to both hard work and good luck. Both Wendi and Rianne cited "luck" as playing a role in their mathematics achievement. Like Rianne, Wendi indicated that when she got a bad grade in mathematics, it was a result of not working hard enough and the teacher not liking her. Wendi and Rianne were the only participants who attributed teacher attitudes and luck to learner success or failure in mathematics. Although Wendi indicated that she utilized multiple methods to verify her answers, like Rianne she stated that an answer was correct if "it just looks right." She elaborated, "I have this feeling [laugh] inside, then if somebody's right there they check it over." Like Rianne, Wendi had a very low mathematics self-concept. Toni recognized these insecurities and provided some insight into how she believed their insecurities influenced their classroom behaviors:

It's almost as if she [Rianne] and Wendi are, have the same problem and are responding in different ways, uh, Rianne is kind of in like this [curls her body and puts her head down] and Wendi responds to her insecurity as being very boisterous so it's, it's, I think it's the same problem there, they've both had humiliating or bad experiences, with it.

Learning Patterns

Wendi was the only one of the learners who stated that she actually "hated" mathematics. Her performance in class fluctuated greatly, ranging from eagerness and enthusiasm to disinterest, to frustration, and even anger. In the weeks before Wendi voluntarily admitted herself into a drug and alcohol rehabilitation program, her classroom behavior was so disruptive that the other learners complained that her actions were interfering with their learning. Despite having received a great deal of one-on-one nurturing and support from preservice teachers, Wendi was unable to overcome her personal problems. As such, she was not always able to participate in the planned mathematical activities and discussions. However, observations of Wendi's learning patterns reveal that, when she was not overwhelmed by personal and emotional problems, she was able to participate and succeed in mathematical activities. She was often confused and even emotionally upset by the open-ended nature of some of the investigative strategies employed by the preservice teachers. Although Wendi agreed that good mathematics teachers would provide learners with a variety of ways to look at a problem, her classroom behavior suggested that she preferred the "security" of a more traditional approach to instruction. This was evident in observations of the positive way in which Wendi responded to a step-by-step, closed-ended, formula-oriented approach to instruction.

Mathematical Development

Wendi's erratic attendance, her disruptive and unfocused classroom behavior, and her inability to complete the course, made it difficult to track her mathematical development over the course of the semester. Her personal problems, as well as her dislike of mathematics, combined to inhibit Wendi's mathematical development. She frequently became extremely agitated during open-ended, exploratory activities. This is revealed in the following transcript excerpt from a lesson on graphing on the life-size Cartesian axes:

Wendi: This is why I hate math, it's so confusing.
Toni: Well.
Mark: [to Wendi] You were getting that stuff.
Toni: This is, this is what we're working with, $y = -x$.
Wendi: But things keep changing, and it's like they're getting more jumbled up from what they are right now.

Geometry and Spatial Sense

Wendi entered the course without any background or knowledge of formal geometry. Because of her erratic attendance, she displayed inconsistent development of her geometric knowledge and understandings. Early in the semester, Wendi attempted to find the area of a rectangle by adding all of the sides. She referred to both a rectangle and a parallelogram as "squares" with different lengths. In a subsequent interview, Wendi referred to both a rectangle and a parallelogram as "squares" but then changed her response to call them both, "rectangles." She also referred to the lengths and widths of the polygons as "area." However, she was able to recognize that the interior angle measure for each figure would be 360°:

> Researcher: Ok, what's the total angle measure for this figure? [points to rectangle]
> Wendi: 360 degrees.
> Researcher: How do you know?
> Wendi: These are all 90degrees right around here. [traces around figure]
> Researcher: Ok, how about this one? [parallelogram]
> Wendi: That would be 360 too. [laughs]
> Researcher: What makes you say that?
> Wendi: Because [pause] no clue but [pause] they just all add up to 360. [laughs]
> Researcher: Are you thinking of the formula?
> Wendi: No. [pause]
> Researcher: What makes you think it's 360?
> Wendi: You're going all around it, like this. [traces around the figure]

Number Sense

Throughout the semester, Wendi displayed limitations in number sense. Her problem solving and understandings of number appear to be heavily influenced by a procedural focus:

> Toni: Sorry, so how do I turn that into a percent?
> Wendi: Move the decimal over to the right 2 times.
> Toni: What are you doing when you do that?
> Jamie: Adding...
> Wendi: Turning it into a percent.
> Toni: Yeah, but what process are you using when you do that?
> Wendi: I don't know.

Wendi's focus on procedures and her dependency on the calculator appear to have contributed to her inability to ascertain the reasonableness of an answer. For example, during an in-class stock market activity, she was asked to find 11/16 of one dollar:

> Melissa: Right, you're going to do like you did before when you had 3/10 and you wanted to represent that as a decimal, you divided the 10 into the 3 and you got point 3, so if you divide 16 into 11, you're going to get a decimal, representation,

and then that way you can figure out how much you can buy [pause, students calculating answers]
Wendi: I got a dollar 45, I got like 1 point 45.
Melissa: For 11 divided by 16, you got what now?
Toni: How can part of a dollar equal more than a dollar?
Wendi: Yeah, I don't know.

Wendi resisted conceptual explanations. During a lesson in which the preservice teachers utilized a rectangular model to teach operations with fractions Wendi explained, "I don't like the visuals."

Functions and Graphing

Wendi indicated that she had taken "algebra 1 first semester" and "intro algebra." She revealed that her experiences in these classes were "horrible" because "nobody ever helped me out with it." As a result, Wendi entered the course with an extremely low algebra self-concept. At midsemester, after having explored functions in class, Wendi was unable to determine or identify a function. Wendi's frustrations were exhibited during a lesson on graphing when she repeatedly stated that she was "lost" and "confused." She demonstrated very little understanding of linear equations and was unable to devise a strategy to solve a problem which differed from those which were presented in class:

Researcher: This next one might be a little different from anything you've seen in class [Solve: $3x + 4 = 9x - 5$]
Wendi: [laughs] This is funny.
Researcher: Why is that?
Wendi: Cause, it just is, it's different.
Researcher: Does any of it look familiar?
Wendi: The $3x + 4$.
Researcher: What's that from?
Wendi: When we were out on the graph. [referring to the life size graph]
Researcher: Ok.
Wendi: But [studies problem]
Researcher: Where would you start?
Wendi: I don't know, no idea.
Researcher: No idea?
Wendi: No.

Effect of the Teaching Experience on the Learners

Engaging in the teaching experience impacted the learners in several ways. Over the course of the semester, there were changes to the learners' self-concepts and mathematical development. Although they impacted self-concept and learning patterns in various ways, for the most part, the learners' beliefs appeared to be unchanged. Like the preservice teachers, it appears that the learners' beliefs were rooted in their prior educational experiences. However, the procedural-focus of many of the mathematics lessons may have served to reinforce the learners' beliefs that success in mathematics was a matter of learning and applying a series of formulas and procedures. Participating in the teaching experience did appear to have a positive effect upon the learners' self-concepts. In turn, the learners' self-concepts appear to have had a positive effect upon their mathematical development. The mutually supportive mathematical learning environment, which encouraged and facilitated positive social interactions, personal bonds, and a feeling of mutual respect and trust to develop between teacher and learner, may have enhanced learner participation. The learners demonstrated mathematical development in several areas and made significant gains in geometry and spatial sense. They also demonstrated proportional reasoning and were able to apply this knowledge in appropriate contexts. The learners displayed functional understandings of correspondence, variable, and slope.

For Exploration and Reflection

1. After reading through the lesson transcripts, did any of the learners' responses surprise you? Why? Why not?
2. What strengths can you identify in each learner? Provide examples to support your response.
3. What adaptations or modifications would you make to your lessons in order to accommodate each of these learners? Why? Clarify and justify your response.
4. Identify examples of learners listening to, responding to, and questioning the teacher and one another during the lessons.
5. What insights did you gain about students on the margin from reading excerpts of these lessons?
6. Based upon the learners' responses, is there anything from these lessons that you plan to use in your own teaching? Explain.
7. What other activities related to these topics could you plan? What mathematical connections could be emphasized? What mathematics might emerge? How might the learners respond?
8. Design a lesson that addresses one of the content areas from this chapter and enables students to communicate and solve real life problems. Include the following elements:

   (a) Objective: detail what students should know about this particular mathematical content area. Why is area important? What do you want your students to be able to do and think? Your objectives should be observable and measurable.

   (b)Philosophy: describe your philosophy for teaching this mathematical content area to students. What teaching strategies and approaches will facilitate students achieving the goals you have identified for them in your Objective section?

   (c) Identify the NCTM and state standards that your lesson addresses.

(d) Prerequisite Knowledge or Prior Learning Experiences: identify the set of knowledge, skills, and understandings that students should bring to this lesson.

(e) List the materials that are required for this lesson: manipulatives, supplies, technology tools, handouts, books

(f) Procedures: Clearly describe what will occur during the course of the lesson. Your lesson should include a description of what will occur during each of the following: Motivating Activity (Anticipatory Set), Developmental Activity (Detail the procedures you will use, including instructional strategies, questions, examples, statements, and descriptions of activities and problems), Culminating Activity (Closure)

(g) Extension: projects, explorations, higher-level thinking questions/applications

(h) Modifications: adaptations for diverse learners. What if the students finish early? The lesson is too easy? Too hard?

9. Create a set of handouts and activity sheets to accompany your lesson.
10. Decide what assessment information your lesson should provide and how it can best be gathered. Your assessment should be authentic and connect to your learning objectives. Create a rubric to assess students' understandings of the mathematics concepts in your lesson.

Chapter Six

Conclusion

This book has taken the reader on a journey with preservice teachers and students in an alternative high school in search of insights into how to foster mathematical success for students on the margin. By providing detailed information and examples of what preservice teachers believe and do in response to a student population composed of marginalized learners this book illustrates how the alternative high school provided a unique environment for the preservice teachers to engage in a collaborative process of reexamining and challenging their conceptions of mathematics teaching and constructs of student knowledge and student learning. This book also provides insight into whether and how conceptually-focused, student-centered teaching can promote changes to preservice teachers' instructional practices and foster mathematics achievement among students who have been labeled "at-risk." As illustrated in the case studies presented here, the pre-student teaching experience has important implications for mathematics teaching and learning. Specifically, that a pre-student teaching experience in an alternative high school can enhance preservice teachers' understandings of how students think mathematically and how to cultivate mathematical thinking in these students.

The pre-student teaching experience examined here is designed to foster and study the development and implementation of enhanced mathematics instructional programs for marginalized and disconnected students who had not flourished in traditional classrooms. Although they brought their own perceptions about mathematics teaching and marginalized learners into the classroom, working with the students assisted the preservice teachers in challenging their beliefs about mathematics, refining their knowledge of mathematics teaching, and reconceptualizing their perceptions of students on the margin. By enabling the preservice teachers to experience the complexities and dynamics of teaching mathematics in an alternative high school, the pre-student teaching experience served as a vehicle for developing the preservice teachers' instructional practices. Through their interactions with the learners, the preservice teachers came to value conceptually-focused, student-centered teaching as an essential element in effective mathematics learning. By viewing the class as a community of learners and creating a supportive environment for students whose prior mathematics learning experiences were negative, the preservice teachers supported the construction of the students' mathematical understanding and achievement

The pre-student teaching initiative attempted to provide ongoing preservice teacher preparation geared for students on the margin; emphasize teaching mathematics through context; encourage the demonstration of the relationships between mathematics and the lives of the students, and, the use of technology tools, collaborative learning groups, and student comments to facilitate discussion. Rather than focusing on the marginalized students' perceived deficits, during their mathematics lessons, the preservice teachers emphasized assisting the learners in constructing positive self-concepts, treating them with respect, and continually affirming and validating their worth. In addition, as they demonstrated changes to their instructional practices, the pre-service teachers made a conscious effort to enhance how these learners experienced mathematics by addressing their backgrounds, interests, and needs. By providing some insight into the complexities of mathematics teaching and learning in an alternative high school, the voices, visions, and perspectives of the preservice teachers and students presented here reveal how conceptually-focused, student-centered teaching can begin to break the cycle of educational failure for marginalized, disconnected, and "at-risk" students.

For Exploration and Reflection

1. Visit several mathematics classrooms and document the problem solving strategies that you observe the students employing. Were the students given time to reflect on and communicate about the problem solving process? What did you learn from observing the students? What student misunderstandings or misconceptions did you observe? How were the students you observed during your classroom visits similar to and different from the learners portrayed in this book?
2. How did the students you observe construct mathematical knowledge in ways that were meaningful to them? In what ways were the students guided and supported as they attempted to generate different strategies? Describe. Identify the process standards that are employed in the mathematics classrooms you observed: Based upon your reflections on your classroom observations, describe what follow-up activities you would design to capitalize on the students' learning.

3. What are the advantages of learning mathematics through problem solving? What are the roles played by the teacher and students in a problem-based learning environment?
4. What evidence do the learners profiled in these pages and the learners you observed during your classroom visits provide to support the notion that all students can do mathematics?
5. What insights did you gain from reading about the preservice teachers and learners in this book? What issues and perspectives related to the mathematics education of students on the margin can you identify? Describe any ideas that you plan to use in your own teaching. Have your own beliefs about mathematics teaching and learning and perceptions of students on the margins changed? Describe.

Bibliography

Anthony, G. (1996). Active learning in a constructivist framework. *Educational Studies in Mathematics*, *31*(4), 349-369.

Ball, D. L. (1990). Breaking with experience in learning to teach mathematics: The role of a preservice methods course. *For the Learning of Mathematics* *10*(2), 10-16.

Baptiste, H. P. (1992). Conceptual and theoretical issues. In H. C. Waxman, J. W. de Felix, J. E. Anderson & H. P. Baptiste (Eds.), *Students at risk in at-risk schools* (pp. 11–16). Newbury Park, CA: Corwin Press.

Bauersfeld, H. (1995). Language games in the mathematics classroom: Their function and their effects. In P. Cobb & H. Bauersfeld (Eds.), *The emergence of mathematical meaning: Interaction in classroom cultures* (pp. 211-292). Hillsdale, NJ: Lawrence Erlbaum.

Belenky, M., F., Clinchy, B. M., Goldberger, N. R., & Tarule, J. M. (1985). *Women's ways of knowing: The development of self, voice, and mind.* New York: Basic Books.

Benninga, J. S., Guskey, T. R.,& Thornburg, K. R. (1981). The relationship between teacher attitudes and student perceptions of classroom climate. *Elementary School Journal*, *82*, 66-75.

Bergeron, J. C., & Herscovics, N. (1990). Psychological aspects of learning early arithmetic. In P. Nesher & I. Kilpatrick (Eds.), *Mathematics and cognition: A research synthesis by the International Group for the Psychology of Mathematics Education.* ICMI study series (pp. 31-52). Cambridge, UK: Cambridge University Press.

Borasi, R. (1990). The invisible hand operating in mathematics instruction: Students' conceptions and expectations. In T. J. Cooney & C. R. Hirsch, (Eds.), *Teaching and learning mathematics in the 1990s (1990 Yearbook).* Reston, VA: National Council of Teachers of Mathematics.

Borasi, R. (1992). *Learning mathematics through inquiry.* Portsmouth, NH: Heinemann.

Borasi, R. (1996). *Reconceiving mathematics instruction: A focus on errors.* Norwood, NJ: Ablex.

Borko, H., Eisenhart, M., Brown, C.A., Underhill, R.G., Jones, D., & Agard, P.C. (1992). Learning to teach hard mathematics: Do novice teachers and their instructors give up too easily? *Journal for Research in Mathematics Education*, *23*, 194-222.

Brophy, J. E. (1985). Teacher-student interaction. In J. B. Dusek (Ed.), *Teacher expectancies.* Hillsdale, NJ: Lawrence Erlbaum.

Brown, J. S., Collins, A., & Duguid, P. (1989). Situated cognition and the culture of learning. *Educational Research, 18*(1), 32-42.

Byers, V. & Herscovics, N. (1977). Understanding school mathematics. *Mathematics Teaching 81*, 24-27.

Byrne, B. M. (1984). The general/academic self-concept nomological network: A review of construct validation research. *Review of Educational Research, 54,* 427-456.

Byrne, B. M., & Shavelson, R. J. (1986). On the structure of adolescent self-concept. *Journal of Educational Psychology, 78*, 474-481.

Cabello, B., & Burstein, N. D. (1995). Examining teachers' beliefs about teaching in culturally diverse classrooms. *Journal of Teacher Education,* 46(4), 285-294.

Calderhead, J. (1984). *Teachers' classroom decision making.* London: Holt, Rinehart.

Campbell, P., & Langrall, C. (1993). Making equity a reality in classrooms. *The Arithmetic Teacher* (10), 110–113.

Carraher, T.N., Schliemann, A.D. & Carraher, D.W. (1988). Mathematical concepts in everyday life. In G. Saxe & M. Gearhart (Eds.) *Children's Mathematics, New Directions for Child Development* (pp. 71-87). San Francisco: Jossey Bass.

Carey, D., Fennema, E., Carpenter, T., & Franke, M. (1995). Equity and mathematics education. In W. G. Secada, E. Fennema, & L. B. Adajian (Eds.), *New directions for equity in mathematics education* (pp. 93–125). New York: Cambridge University Press.

Carpenter, T. P., Fennema, E., Petersen, P., & Carey, D. (1988). Teachers' pedagogical content knowledge of students' problem solving in elementary arithmetic. *Journal for Research in Mathematics Education, 19*, 385-401.

Cazden, C. (1986). Classroom discourse. In M. Wittrock (Ed.), *Handbook of research on teaching* (3rd ed., pp. 432-463). New York: Macmillan.

Cazden, C.B., John, V.P., & Hymes, D. (Eds.). (1972). *Functions of language in the classroom.* New York: Teachers College Press.

Chaiklin, S. (1989) Cognitive studies of algebra problem solving and learning. In S. Wagner & C. Kieran (Eds.) *Research issues in the learning and teaching of algebra* (pp. 93-114). Reston, VA: National Council of Teachers of Mathematics and Lawrence Erlbaum Associates.

Cobb, P. (1994). Where is the mind? Constructivist and sociocultural perspectives on mathematical development. *Educational Researcher*, 23(7), 13-20.

Cobb, P. (1995). Cultural tools and mathematical learning: A case study. *Journal for research in mathematics education*, *26*, 362-385.

Cobb, P., & Wheatley, G. (1988). Children's initial understandings of ten. *Focus on Learning Problems in Mathematics.* 10 (3), 1-28.

Cobb, P., Yackel, E., & Wood, T. (1991). Curriculum and teacher development. In E. Fennema, T. P. Carpenter, & S. J. Lamon (Eds.), *Integrating research on teaching and learning mathematics* (pp. 153-176). Albany, NY: State University of New York Press.

Comer, J. P. (1988). Educating Poor Minority Children. *Scientific American* 259 (5), 42-48.

Conference Board of the Mathematical Sciences (2001). *The mathematical education of teachers.* Washington, DC: Mathematical Association of America and American Mathematical Society.

Cooney, T. J. (1985). A beginning teacher's view of problem solving. *Journal for Research in Mathematics Education*, 16(5), 324-336.

Cooney, T. J., & Wilson, M. R. (1993). Teachers' thinking about function: Historical and research perspectives. In T. Romberg, E. Fennema, & T. Carpenter (Eds.), *Integrating research on the graphical representation of function* (pp. 131-158). Hillsdale, NJ: Lawrence Erlbaum.

Crawford, K. P. (1992). Applying theory in teacher education: Changing practice in mathematics education. In W. Geeslin and K. Graham (Eds.), *Proceedings of the Sixteenth PME Conference* (pp. 161-167). Durham, NH: University of New Hampshire.

Cummins, J. (1989). Language and Literacy Acquisition in Bilingual Contexts. *Journal of Multilingual and Multicultural Development* 19: 17-31.

Davis, R. B. (1984). *Learning Mathematics: The Cognitive Science Approach to Mathematics Education.* Norwood, NJ: Ablex Publishing.

Davis, R.B. (1992). Reflections on where mathematics education now stands and on where it may be going. In D. A. Grouws, (Ed.) *Handbook of research on mathematics teaching and learning* (pp. 724-734). New York: MacMillan.

Davis, P. J., & Hersh, R. (1987). *Descartes' dream: The world according to mathematics*. New York: Houghton Mifflin.

Davis, R. B., Maher, C. A., & Noddings, N. *Constructivist views on the teaching and learning of mathematics* Journal for research in mathematics education. Monograph ; no. 4. Reston, Va.: National Council of Teachers of Mathematics, 1990.

Dorfler, W. (1995). Mathematical objects, representations, and imagery in R. Sutherland & J. Mason (Eds.) *Exploiting mental imagery with computer in mathematics education.* Nato Asi Series F, Vol. 138. Berlin: Springer Verlag, 82-94

Dossey, J. (1992). The nature of mathematics: its role and influence. In D.A. Grouws (Ed.), *Handbook of research on mathematics teaching and learning.* New York: MacMillan.

Dunn, T. K. (2001). *The Effects of an Exploratory Mathematics Course on the Self-Concepts and Mathematical Development of Secondary Students Placed at Risk.* Annual Meeting of the American Educational Research Association, Seattle, WA.

Dunn, T. K. (2001). *Teachers' Mathematics-Related Knowledge, Instructional Practice, and Student Mathematical Development: A Transnational Study of Novice Mathematics Teachers and the Function Concept.* Annual Meeting of the American Educational Research Association, Seattle, WA.

Dunn, T. K. (2002). Teaching Mathematics in the Spirit of Education Reform: Perspectives from the United States and Namibia. In D. S. Mewborn, P. Sztajn, D. Y. White, H. G. Wiegel, R. L. Bryant, and K. Nooney (Eds.), *Proceedings of the Twenty-Fourth Annual Meeting of the North American*

Chapter of the International Group for the Psychology of Mathematics Education. Columbus, OH: ERIC Clearinghouse for Science, Mathematics, and Environmental Education.

Eisenhart, M., Borko, H., Underhill, R.G., Brown, C.A., Jones, D. & Agard, P. (1993). Conceptual knowledge falls through the cracks: Complexities of learning to teach mathematics for understanding. *Journal for Research in Mathematics Education, 24*, 8-40.

Erlwanger, S. H. (1975). Case studies of children's conceptions of mathematics. *Journal of Children's Mathematical Behaviour, 1*(3), 157-283.

Ernest, P. (1989). The Impact of Beliefs on the Teaching of Mathematics. In P. Ernest (Ed.) *Mathematics teaching: The state of the art*, London, Falmer Press, 249-254.

Ernest, P. (1991). *The philosophy of mathematics education.* London: Falmer.

Eubanks, E. E., & Parish, R. (1992). Leadership that promotes instability: A hope for at-risk students. In H. C. Waxman, J. Walker de Felix, J. E. Anderson, & H. P. Baptiste (Eds.), *Students at risk in at-risk schools* (pp. 143-159). Newbury Park, CA: Corwin Press.

Feiman-Nemser, S., & Parker, M. B. (1990). Making subject matter part of the conversation in learning to teach. *Journal of Teacher Education, 41*, 32-43.

Fennema, E., & Franke, M. L. (1992). Teachers' knowledge and its impact. In D. A. Grouws (Ed.), *Handbook of research on mathematics teaching and learning* (pp. 147-164). New York: Macmillan Publishing Company.

Fennema, E., & Peterson, P. (1985). Autonomous learning behavior: A possible explanation of gender-related differences in mathematics. In L. C. Wilkinson & C. B. Marrett (Eds.), *Gender-related Differences in Classroom Interactions* (pp. 17-35). New York: Academic Press.

Fennema, E., & Meyer, M. R. (1989). Gender, equity, and mathematics. In W. Secada (Ed.), *Equity in education* (pp. 146-57). London: Falmer.

Filloy, E. & Rojano, T. (1989) Solving equations: The transition from arithmetic to algebra. *For the Learning of Mathematics*, *9*(2), 19-25

Forman, E.A. & Cazden, C.B. (1985). Exploring Vygotskian perspectives in education: The cognitive value of peer interaction. In J.V. Wertsch (Ed.) *Culture, communication and cognition: Vygotskian perspectives* (pp. 323–347). Cambridge: Cambridge University Press.

Glaser, B. G., & Strauss, A. L. (1967). *The discovery of grounded theory: Strategies for qualitative research.* Chicago: Aldine.

Goetz, J. P., & LeCompte, M. D. (1984). *Ethnography and qualitative design in educational research.* New York: Academic Press.

Goldenberg, C. (1992). Instructional conversations: Promoting comprehension through discussion. *The Reading Teacher*, 46, 316-326.

Goldin, G. (1990), Epistemology, Constructivism, and Discovery Learning Mathematics, in Davis, R. & Maher, C. & Noddings, N. (Eds.), *Constructivist Views on the Teaching and Learning of Mathematics.* New York: Norton.

Goldin, G. A. & Kaput, J. J. (1996). A joint perspective on the idea of representation in learning and doing mathematics. In L. P. Steffe & P. Nesher (Eds.), *Theories of Mathematical Learning* (pp. 397-430). Mahwah, NJ: Lawrence Erlbaum Associates.

Grant, C.A., & Sleeter, C.E. (1993). Race, class, gender, exceptionality, and educational reform. In G.A. Banks & C.A.M. Banks (Eds.), *Multicultural education: Issues and perspectives* (2nd ed., pp. 48-68). Boston: Allyn and Bacon.

Gray, E. & Tall, D. O. (1994). Duality, ambiguity, and flexibility. In *Journal for Research in Mathematics Education, 25* (2), 116-140.

Grossman, P.L. 1990. *The making of teaching: teacher knowledge and teacher education.* New York: Teachers College Press.

Grossman, P. L., Wilson, S. M., & Shulman, L. (1989). Teachers of substance: Subject matter knowledge for teaching. In M. C. Reynolds (Ed.). *Knowledge base for the beginning teacher* (pp. 23-36). Oxford: Pergamon Press.

Halliday, M.A.K. (1975). *Learning how to mean.* London, U.K.: Edward Arnold.

Hatano, G. (1988). Social and motivational bases for understanding. In G. B. Saxe & M. Gearhart (Eds.), *Children's mathematics* (pp. 55-70). San Francisco: Jossey-Bass.

Hatano, G. (1996). A conception of knowledge acquisition and its implications for mathematics education. In L. P. Steffe & P. Nesher (Eds.) *Theories of Mathematical Learning* (pp. 197-217). Hillsdale, NJ: Lawrence Erlbaum Associates.

Herscovics, N. (1989). Cognitive obstacles encountered in the learning of algebra. In S. Wagner & C. Kieran (Eds.), *Research issues in the learning and teaching of algebra.* Reston, VA: National Council of Teachers of Mathematics.

Herscovics, N., & Linchevski, L. (1994). A cognitive gap between arithmetic and algebra. *Educational Studies in Mathematics*, *27*, 59-78.

Hiebert, J., & Carpenter, T. P. (1992). Learning and teaching with understanding. In D. Grouws (Ed.), *Handbook for research on mathematics teaching and learning* (pp. 65-97). New York: MacMillan

Hiebert, J., & Lefevre, P. (1986). Conceptual and procedural knowledge in mathematics: An introductory analysis. In J. Hiebert (Ed.), *Conceptual and procedural knowledge: The case of mathematics* (pp. 1-27). Hillsdale, NJ: Lawrence Erlbaum Associates.

Hiebert, J. & Wearne, D. (1986). Procedures over concepts: The acquisition of decimal number knowledge. In J. Hiebert (Ed.), *Conceptual and procedural knowledge: The case of mathematics* (pp. 199-223). Hillsdale, NJ: Lawrence Erlbaum Associates.

Hiebert, J. & Wearne, D. (1992). Links between teaching and learning place value with understanding in first grade. *Journal for Research in Mathematics Education*, *23*(2) 98-122.

Hiebert, J., & Wearne, D. (1993). Instructional tasks, classroom discourse in students in second grade arithmetic. *American Educational Research Journal, 301,* 393-425.

Jones, B. F. (1988). Toward redefining models of curriculum and instruction for students at-risk. In B. Z. Presseisen (Ed.), *At-risk students and thinking: Perspectives from research.* Washington, D. C: National Education Association.

Kagan, D.M. (1992). Professional growth among pre-service teachers and beginning teachers. *Review of Educational Research, 62*(2), 129-169.

Kaput, J., & Sims-Knight, J. E. (1983). Exploring difficulties in transforming between natural language and image-based representations and abstract symbol systems of mathematics. In D. Rogers & J. A. Sloboda (Eds.), *The acquisition of symbolic skills*. New York: Plenum.

Kaput, J., & Sims-Knight, J. E. (1983). Misconceptions of algebraic symbols: Representations and component processes. In H. Helm & J. D. Novak (Eds.), *Proceedings of the international seminar: Misconceptions in science and mathematics*. Ithaca, NY: Cornell University.

Kieran, C. (1992). The learning and teaching of school algebra, in D. A. Grouws, (Ed.), *Handbook of Research on Mathematics Teaching and Learning*, (pp. 390-419), New York: Macmillan

Kieren, T. (1994). Play and mathematical understanding. *Journal for Research in Childhood Education, 8,* 132-141.

Kilpatrick, J. (1985). Reflection and Recursion. *Educational Studies in Mathematics 16*, 1-26.

Kozol, J. (1991). *Savage inequalities. Children in America's schools*. Crown Publishers, Inc., New York.

Lampert, M. (1986). Knowing, doing, and teaching multiplication. *Cognition and Instruction, 3*, 305-342.

Lampert, M. (1990). When the problem is not the question and the solution is not the answer: Mathematical knowing and teaching. *American Educational Research Journal, 27,* 29-63.

Lappan, G. & Briars, D. (1995). How should mathematics be taught? In I. Carl (Ed.), *Prospects for school mathematics*. Reston, VA: NCTM.

Lave, J. (1991). Situated Learning in Communities of Practice. In L. Resnick, J. Levine & S. Teasley (Eds.), *Perspectives on Socially Shared Cognition.* (pp. 63-82). Washington, D.C.: American Psychological Association.

Lave, J., & Wenger, E. (1991). *Situated learning: Legitimate peripheral participation.* Cambridge, England: Cambridge University Press.

Lemut, E. du Boulay, B., & G. Dettori (1993), Cognitive models and intelligent environments for learning programming (pp. 145-155). Berlin: Springer-Verlag.

Lerman, S. (1989). Constructivism, mathematics and mathematics education. *Educational Studies in Mathematics, 20*, 211 - 223.

Lerman, S. (1990). Alternative perspectives of the nature of mathematics and their influence on the teaching of mathematics. *British Educational Research Journal, 16,* 53 - 61.

Lesh, R., Landau, M., & Hamilton, E. (1983). Conceptual models and applied mathematical problem-solving research. In R. Lesh & M. Landau (Eds.), *Acquisition of mathematics concepts and processes* (pp. 263-344). Orlando, FL: Academic Press.

Levin, H. (1987). *New Schools for the Disadvantaged.* Aurora CO: Mid-Continent Regional Educational Laboratory, 1987.

Levin, H. (1992). *Accelerated Schools for At-risk Students.* New Bruinswick, N.J.: Center for Policy Research in Education, Rutgers University.

Linchevski, L., & Herscovics, N. (1996). Crossing the cognitive gap between arithmetic and algebra: Operating on the unknown in the context of equations. *Educational Studies in Mathematics, 30*(1), 39-65.

Marsh, H.W. (1986). Verbal and math self-concepts: An internal/external frame of reference model. *American Educational Research Journal*, 23, 129-149.

Marsh, H.W. (1987). The big-fish-little-pond effect on academic self-concept. *Journal of Educational Psychology,* 79, 280-295.

Marsh, H.W. (1990). Influences of internal and external frames of reference on the formation of math and English self-concepts. *Journal of Educational Psychology*, 82, 107-116.

Maturana, H. & Valera, F. (1987). *The Tree of Knowledge: The Biological Roots of Human Understanding.* Boston, New Science Library.

Mathematical Sciences Education Board and National Research Council (1990). *Reshaping school mathematics: A framework and philosophy for curriculum.* Washington, DC: National Academy Press.

Matz, M. (1980). Towards a computational theory of algebraic competence, *Journal of Mathematical Behavior, 3* (1), 93–66.

McCown, R., Driscoll, M., & Roop, P.G. (1996). Individual variability in the classroom. *Educational psychology: A learning-centered approach to classroom practice* (2nd ed.): 152-165. Massachusetts; Simon & Schuster Company.

McKnight, C. C., Crosswhite, F. J., Dossey, J. A., Kifer, E., Swafford, J. O., Travers, K. J., & Cooney, T. J. (1987). *The underachieving curriculum: Assessing U.S. school mathematics from an international perspective.* Champaign, IL: Stipes.

McLeod, D.A. (1992). Research on affect in mathematics education: a reconceptualization. In D.A. Grouws (Ed.), *Handbook of research on mathematics teaching and learning* (pp.575-96).New York: Macmillan.

Mead, G. H. (1934). *Mind, Self, and Society.* Chicago: University of Chicago Press

Mehan, H. (1991). *Sociological foundations supporting the study of cultural diversity.* The National Center for Research on Cultural Diversity and Second Language Learning, Research Report 1. Santa Cruz, CA: University of California.

Midgley, C., Feldlaufer, H., Eccles, J. (1989). Change in Teacher Efficacy and Student Self- and Task-Related Beliefs in Mathematics during the Transition to Junior High School. APA, 81(2), 247-258.

Midgley, C., Feldlaufer, H., Eccles, J. (1989). Student/Teacher Relations and Attitudes Toward Mathematics Before and After the Transition to Junior High School. *Child Development*, 60, 981-992.

Miles, M. B., & Huberman, A. M. (1994). *Qualitative data analysis: A sourcebook of new methods*. (2nd ed.). Thousand Oaks, CA: Sage Publications.

Moll, L. C., and Diaz, S. (1987). Change as the goal of educational research. *Anthropology and Education Quarterly*, 18, 300-311.

National Research Council (1989). *Everybody Counts: A Report to the Nation on the Future of Mathematics Education.* Washington, DC: National Research Council.

National Research Council. (1990). *Reshaping school mathematics: A philosophy and framework for curriculum*. Washington, DC: National Academy of Sciences.

National Council of Teachers of Mathematics: Commission on Standards for School Mathematics. (1989). *Curriculum and evaluation standards for school mathematics*. Reston, VA: Author.

National Council of Teachers of Mathematics (1989). *Professional standards for teaching mathematics.* Reston, VA: Author.

National Council of Teachers of Mathematics (2000). *Principles and standards for school mathematics.* Reston, VA: Author.

Novak, J. D., & Gowin, R. (1984). *Learning how to learn.* Cambridge: Cambridge University Press.

Oakes, J. (1985). *Keeping Track: How Schools Structure Inequality*. New Haven: Yale University Press.

Oakes, J. (1986). Tracking, inequality, and the rhetoric of school reform: Why schools don't change. *Journal of Education, 168*, 61-80.

Oakes, J., Ornseth, T., Bell, R., & Camp, P. (1990). *Multiplying inequalities: The effects of race, social class, and tracking on opportunities to learn mathematics and science.* Santa Monica, CA: Rand.

Ogden, E.H., & Germinario, V. (1988). *The at-risk student: Answers for educators*. Lancaster, PA: Technomic Publishing Company.

Pallas, A.M., Natriello, G., & McDill, E. (1989). The changing nature of the disadvantaged population: Current dimensions and future trends. *Educational Researcher*, 18(5), 16- 22.

Palmer, S. E. (1977). Hierarchical structure in perceptual representation. *Cognitive Psychology*, 9, 441 – 474

Patton, M. Q. (1990). *Qualitative evaluation and research methods* (2nd ed.). Newbury Park, CA: Sage.

Peirce, C. S. (1935). *Collected Papers.* Cambridge, MA.: Harvard University Press.

Pimm, D. (1987). *Speaking mathematically.* London, U.K.: Routledge.

Pirie, S. E. B. (1998). Crossing the gulf between thought and symbol: Language as stepping-stones. In H. Steinbring & M. G. B. Bussi & A. Sierpinska (Eds.), *Language and communication in the mathematics classroom.* Reston, VA: National Council of Teachers of Mathematics.

Post, T., Harel, G., Behr, M. & Lesh, R. (1991). Intermediate teachers' knowledge of rational number concepts. In E. Fennema, T. P. Carpenter, & S. J. Lamon (Eds.) *Integrating Research on Teaching and Learning Mathematics.* Albany, New York: SUNY Press, pp. 177-198.

Relich, J. (1996). Gender, Self-Concept and Teachers of Mathematics: Effects on Attitudes to Teaching and Learning. *Educational Studies in Mathematics* 30: 179-195.

Rosenthal, R., & Jacobson, L. (1968). *Pygmalion in the Classroom* New York; Holt, Rinehart & Winston.

Resnick, L. (1995). From aptitude to effort: A new foundation for our schools. *Daedalus, 124*(4), 55–62.

Reyes, L.H. (1984). Affective variables and mathematics education. *The Elementary School Journal*, 84, 558-581.

Rist, R. (1970). Student Social Class and Teacher Expectations: The Self-Fulfilling Prophecy in Ghetto Education. *Harvard Educational Review 40,* 3, 411-451.

Rogoff, B.. (1990). *Apprenticeship in Thinking. Cognitive Development in Social Context.* New York: Oxford University Press.

Schoenfeld, A. (1985). *Mathematical Problem Solving.* New York: Academic Press.

Schoenfeld, A. (1987). *Cognitive Science and Mathematics Education.* Hillsdale, NJ: Erlbaum Associates.

Secada, W. G. (1989). Educational equity versus equality of education: An alternative conception. In W. G. Secada (Ed.), *Equity in education* (pp. 68-88). London: Falmer.

Secada, W. G. (1992). Race, ethnicity, social class, language, and achievement in mathematics. In D. A. Grouws (Ed.), *Handbook of research on mathematics teaching and learning* (pp. 623-660). New York: Macmillan.

Sfard, A. (1991). On the dual nature of mathematical conceptions: Reflections on processes and objects as different sides of the same coin. *Educational Studies in Mathematics, 22*, 1-36.

Sfard, A. (1994). Reification as the birth of metaphor. *For the learning of mathematics, 14*(1), 44-55.

Sfard, A. (1995). The development of algebra: Confronting historical and psychological perspectives. *Journal of Mathematical Behavior, 14,* 15-39.

Shavelson, R., Hubner, H., & Stanton, G. (1976). Self-concept: Validation of construct interpretations. *Review of Educational Research,* 46, 407-441.

Shavelson, R., & Bolus, R. (1982). Self-concept: The interplay of theory and methods. *Journal of Educational Psychology,* 74, 3-17.

Shulman, L. (1986). Those who understand: Knowledge growth in teaching. *Educational Researcher, 15* (2), 4-14.

Shulman, L. S. (1987). Knowledge and teaching: Foundations of the new reform. *Harvard Educational Review 57*(1), 1-22.

Siegel, M., & R.F. Carey. (1989). *Critical thinking: A semiotic perspective.* Bloomington, IN: ERIC Clearinghouse on Reading and Communication Skills.

Silver E. A., Smith, M. S., & Nelson, B. S. (1995). The QUASAR Project: Equity concerns meet mathematics education reform in the middle school. In W. G. Secada, E. Fennema, & L. B. Adajian (Eds.), *New directions for equity in mathematics education* (pp. 9–56). New York: Cambridge University Press.

Sims-Knight, J. & Kaput, J. J. (1983). Exploring difficulties in transformations between natural language and image based representations and abstract symbol systems of mathematics. In D. Rogers & J. Sloboda (Eds.), *The acquisition of symbolic skills* (pp. 561-569). New York: Plenum.

Sinclair, R.L., & Ghory, W.J. (1987). *Reaching marginal students: A primary concern for school renewal.* Berkley, CA: McCutchan Publishing Corporation.

Slavin, R.E. (1989). Students at risk of school failure: The problem and its dimensions. In R.E. Slavin, N.L. Karweit, & N.A. Madden (Eds.), *Effective programs for students at-risk* (pp. 3-17). Needham Heights, MA: Allyn & Bacon.

Sleeter, C. E. & Grant, C. A. (1988). A Rationale for Integrating Race, Gender, and Social Class. In L. Weis (Ed.) *Class, Race, and Gender in American Education.* (pp. 144-160). New York: State University of New York Press.

Spindler, G. D. (1982). *Doing the ethnography of schooling.* New York: Holt, Rinehart, and Winston.

Spradley, J. P. (1979). *The ethnographic interview.* New York: Holt, Rinehart, & Winston.

Stanic, G. M. A & Hart, L. E. (1995). Attitudes, persistence, and mathematics achievement: Qualifying race and sex differences. In Secada, W. G. (Ed.). *New Directions for Equity in Mathematics Education.*, pp. 242-257. Cambridge, MA: Cambridge University Press.

Steffe, L. P. & Wiegel, H. (1996). On the nature of a model of mathematical learning. In L. P. Steffe & P. Nesher (General Editors), *Theories of mathematical learning (pp. 477-498).* Hillsdale, NJ: Lawrence Erlbaum Associates.

Strauss, A. L. (1987). *Qualitative analysis for social scientists.* New York: Cambridge University Press.

Strauss, A. L., & Corbin, J. (1990). *Basics of qualitative research: Grounded theory procedures and techniques.* Newbury Park, CA: Sage.

Strutchens, M. E. (2000). Confronting beliefs and stereotypes that impede the mathematical empowerment of African American students. In M. E. Strutchens, M. Johnson, & W. Tate (Eds.). *Changing the faces of mathematics: Perspectives on African Americans.* (pp. 7-14). Reston, VA: National Council of Teachers of Mathematics.

Strutchens, M. E., & Silver, E. A. (2000). NAEP findings regarding race/ethnicity: The students, their performance, and their classrooms. In E. A. Silver & P. A. Kenney (Eds.), *Results from the seventh mathematics assessment of the National Assessment of Educational Progress* (pp. 45-72). Reston, VA: National Council of Teachers of Mathematics.

Tate, W.F. (1995). Returning to the root: A culturally relevant approach to mathematics pedagogy. *Theory Into Practice, 34*, 166-173.

Tharp, R. G., & Gallimore, R. (1988). *Rousing minds to life: Teaching, learning, and schooling in social context.* New York: Cambridge University Press.

Tharp, R. G., & Gallimore, R. (1989). Rousing schools to life. *American Educator,* 13(2),20- 25,46-52.

Thompson, A. G. (1984). The relationship of teachers' conceptions of mathematics and mathematics teaching to instructional practice. *Educational Studies in Mathematics, 15*(2), 105-27.

Thompson, A. G. (1992). Teachers' beliefs and conceptions: A synthesis of research. In D. A. Grouws (Ed.), *Handbook of research on mathematics teaching and learning* (pp. 127-146). New York: Macmillan.

Tobias, R. (1992). Raising student self-esteem through mathematics and science. In R. Tobias (Ed.), *Nurturing at-risk youth in math and science: Curriculum and considerations* (pp. 103-120). Bloomington, Indiana: National Educational Service.

Vatter, T. (1992). Teaching Mathematics to the At-Risk Secondary School Student. *Mathematics Teacher 85*(4), 292-94.

von Glasersfeld, E. (1990) Environment and communication. In L. P. Steffe & T. Woods (Eds.), *Transforming children's mathematical education* (pp. 30-38). Hillsdale, N.J.: Erlbaum.

von Glasersfeld, E.: (1987). Learning as constructive activity, in *The Construction of Knowledge Contributions to Conceptual Semantics* (pp. 307-333). Seaside, CA: Intersystems Publications.

Vygotsky, L. S. (1978). *Mind in Society. The Development of Higher Psychological Processes*. Cambridge, MA: Harvard University Press

Walker de Felix, J., Baptiste, Jr., H. P., and Anderson, J. (1992). Future directions for educating students at risk. In H. C. Waxman, J. Walker de Felix, J. E. Anderson, & H. P. Baptiste, Jr., (Eds.), *Students at risk in at-risk schools*. Los Angeles: Corwin.

Waxman, H. C. (1992). Reversing the cycle of educational failure for students in at-risk school environments. In H. C. Waxman, J. Walker de Felix, J. Anderson, & H. P. Baptiste (Eds.), *Students at risk in at-risk schools: Improving environments for learning* (pp. 1-9). Newbury Park, CA: Sage.

Wearne, D., & Hiebert, J. (1988). A cognitive approach to meaningful mathematics instruction: Testing a local theory using decimal numbers. *Journal for Research in Mathematics Education, 19*, 371-384.

Weiner, B. (1986). *An attributional theory of motivation and emotion.* New York: Springer-Verlag.

Wilson, S. M., Shulman, L. S., & Richert, A. E. (1987). 150 different ways of knowing: Representation of knowledge in teaching. In J. Calderhead (Ed.). *Exploring teachers' thinking* (pp. 104-124). London: Cassell.

Index

alternative high school, 7, 24
attitudes, 16
attitudes, teacher, 17-18
attitudes, learners, 18-19
at-risk learners, 2-7
backgrounds, learners', 36-37
backgrounds, preservice teachers', 31-36
beliefs and attitudes, 16
beliefs and attitudes, learners, 18-19, 106-108
beliefs, preservice teachers, 37-40'
beliefs and attitudes, teacher, 17-18
case studies, preservice teachers, 61-103
case studies, learners, 113-145
collaborative learning groups, 56-57
conceptual knowledge, 9
connecting knowledge to symbols, 11-12
context, 24
data analysis, 25-28
data collection, 25
discourse, mathematical, 20-21
discussion, facilitating, 57-61
environment, educational, 4-5
environment, learner's, 5-6
equity, 1, 2
functions and graphing, 117, 125, 131, 138, 145
geometry and spatial sense, 115, 123, 128, 135, 144
graphing, 117, 125, 131, 138, 145
instructional practices, 2, 26
instructional practices, preservice teachers, 42-61
knowledge, connecting to symbols, 11-12
knowledge, teacher, 15-16
knowledge in context, 10-11
learners, 36-37
learning patterns, 115, 122, 127, 134, 143
marginalized learners, 2-7
mathematical development, learners', 109-113
mathematical discourse, 20-21
mathematical knowledge, 8-9
mathematical power, 2
mathematical self-concept, 19
mathematical understanding, 13-15
mathematics education, 1
mathematics education reform, 1-2
mathematics lessons,
number sense, 117, 124, 130, 137, 144
participants, 24
preservice teachers, 31-36
preservice teachers, instructional practices, 42-61
preservice teachers, perceptions of students on the margin, 40-42
pre-student teaching experience, 7
problem solving, 2-5
procedural knowledge, 9
procedural knowledge, teaching for, 42-54
reflection, 14
relationship, conceptual and procedural knowledge, 10
representations, 12-13
self-concept, learners, 108'
self-concept, mathematical, 19
spatial sense, 25, 115, 123, 128, 135, 144
student comments, 57-61
students on the margin, 4-6
teacher attitudes, 3
teacher knowledge, 15-16
teacher preparation, 3-4

teaching conceptually, 54-61
teaching for procedural knowledge, 42-54
technology tools, 54-56
tracking, 3
transforming teaching and learning, 6-7
understanding, mathematical, 13-15

www.ingramcontent.com/pod-product-compliance
Ingram Content Group UK Ltd.
Pitfield, Milton Keynes, MK11 3LW, UK
UKHW041141300926
15471UKWH00011B/93

9 780761 833338